The Art-Science Symbiosis

AF715824

Marcelo Velasco · Ignacio Nieto

The Art-Science Symbiosis

Springer

Marcelo Velasco
Fundación Arte+Ciencia
Santiago, Chile

Ignacio Nieto
Facultad de Artes Visuales
Finis Terrae University
Santiago, Chile

ISBN 978-3-031-47406-4 ISBN 978-3-031-47404-0 (eBook)
https://doi.org/10.1007/978-3-031-47404-0

© The Editor(s) (if applicable) and The Author(s), under exclusive license to Springer Nature Switzerland AG 2024

This work is subject to copyright. All rights are solely and exclusively licensed by the Publisher, whether the whole or part of the material is concerned, specifically the rights of translation, reprinting, reuse of illustrations, recitation, broadcasting, reproduction on microfilms or in any other physical way, and transmission or information storage and retrieval, electronic adaptation, computer software, or by similar or dissimilar methodology now known or hereafter developed.
The use of general descriptive names, registered names, trademarks, service marks, etc. in this publication does not imply, even in the absence of a specific statement, that such names are exempt from the relevant protective laws and regulations and therefore free for general use.
The publisher, the authors and the editors are safe to assume that the advice and information in this book are believed to be true and accurate at the date of publication. Neither the publisher nor the authors or the editors give a warranty, expressed or implied, with respect to the material contained herein or for any errors or omissions that may have been made. The publisher remains neutral with regard to jurisdictional claims in published maps and institutional affiliations.

This Springer imprint is published by the registered company Springer Nature Switzerland AG
The registered company address is: Gewerbestrasse 11, 6330 Cham, Switzerland

Paper in this product is recyclable.

Cover Illustartion: Francois-Joseph Lapoint. 2016. From the series *Microbiome Selfie. 1000 Handshakes*. This image exhibits the extensivemicrobiome resulting from the analysis of the samples collected after 650 handshakes. ArtLaboratory. Berlin, Germany © François-Joseph Lapointe, 2016.

"1000 Handshakes" is a performance piece during which the artist shakes hands with as many people as possible, gradually changingthe invisible microbial community in the palm of his hand. Periodically, assistants take a swab from the skin, and the samples areanalysed in the lab to reveal how our contact with others shapes the microbes between us. This ongoing project has been performedin different cities around the world (including Copenhagen, Montreal, San Francisco, Perth, Berlin, Baltimore and Paris) as a way tomap our collective microbiome using scientific data. Production of the "microbiome selfies" involved many different steps. Following thecollection of the microbiome samples, bacterial DNA was extracted, amplified and sequenced to create the bioinformatic data shown inthis series. The nodes of the networks represent bacterial DNA sequences, and two nodes are connected by an edge (line) when thebacterial DNA sequences have more than 95% similarity. The different colours correspond to distinct samples collected at every 50th handshake, from 0 to 1001.

To Ivan, Sonia, Lucia, Chony, Pato, Andrés, Paula, Reinaldo and Pascuala, my beloved family.

—Marcelo Velasco

To my mother Carmen Luz, who gave me the opportunity to write it down.

—Ignacio Nieto

Foreword

Especially since the beginning of the twenty-first century, there has been a broad debate about how art and science (and especially artists and scientists) should enter into dialogue and constructively engage with each other. Historically, this has been a problematic process, fostered by a genuine interest in structurally different approaches and languages and at the same time systematically burdened by the temptation to subordinate the respective counterpart. More generally, I could argue that the difficult dialogue between the protagonists of science and the humanities, where the respective boundaries have often created a counterproductive and insurmountable intellectual competition, is somehow sublimated.

Velasco and Nieto are a rare example of systematic intellectuals who unite science and art in their writings and practices and have been engaged in such a conversation for years, which has been gradually transformed into a methodological approach in this book. Both operate in Santiago de Chile, geographically on the fringes of the contemporary art market and major science funders, and yet have deeply probed original and radical intersections of art and science that are reflected in as many artworks.

The book seems informed by their conversations and curatorial stance. The authors co-authored the first two chapters, which provide a comprehensive overview of the founding principles they articulated. They then selected 22 artworks that represent, on different levels, the integration of art and science in a single work. They all reflect an original balance between the two fields and involve a range of scientific disciplines. The protagonists' descriptions of the projects are also linked to an analysis of their insights.

This work articulates in itself what the authors define as "a third way in the creation of knowledge" and proves how the methodology applied can be successfully applied to a variety of scientific fields without violating its principles.

Dialogue, as one of the supreme tools that we still have at our disposal to collectively and effectively discuss the big questions of our existence, requires constant negotiation, or what the authors brilliantly define as 'inter-influence'. The precarious 'symmetry' in the selected works and the effort to maintain an inherent balance forms a cultural value in itself, which is consolidated through the same selection.

It is hoped that the dialogue developed by the authors will continue through the artworks and eventually extend to the readers, especially those involved in these fields, to start a virtuous process of cultural interinfluence.

Dr. Alessandro Ludovico
Associate Professor in Art Design
and Media, Winchester School of Art
University of Southampton
Southampton UK

Chief editor,
Neural magazine www.neural.it

Preface

Many of the ideas presented in this book come from our own practice as mediators and promoters of art-science works, mainly among scientists. We learned a great deal from the talented art-scientists involved in the project, from their experiences and artworks. Choosing and obtaining the relevant images allowed us to contact people as interesting and generous as they are enthusiastic.

More than an attempt to provide answers to very specific questions, this book presents the diversity and breadth of art-science integration and offers a framework to analyse these and promote the wider discussion of ways this integration might contribute to form new types of knowledge. While the proposed framework does provide limits and definition to the diversity of art-science works, it is intended more as a heuristic tool than a rigid structure, as we acknowledge having left out some elements and fields such as sound art, architecture and design, each with a robust tradition of its own. The rewarding experience of communication with the talented exponents of art-science who participated in this book, nonetheless, confirms our belief in the value of dialogue among artists and scientists.

This book was written without financial support nor ties of affiliation, a circumstance which gave us independence and freed us from external pressures and institutional purposes.

The Art-Science Symbiosis is motivated by our own personal experiences in and curiosity about art-science integration to which we come equipped with somewhat different perspectives. One of us, Marcelo Velasco, is rooted in Neuroscience, Ecological Economics and Medial Arts, while the other, Ignacio Nieto, is more at home in the Visual Arts, Education and Aesthetics. The reader will easily find signs of these disciplinary perspectives in the pages of this book.

The book was written in Santiago-Chile, in a friendly language and intended for scientists, but without a doubt it can be useful for artists, postgraduate students in science or humanities or by anyone interested in the subject. We would be delighted to be contacted by the readers to continue this discussion with all those who are interested in learning about and developing outstanding art-science challenges. Website: www.artscience-symbiosis.net

Santiago, Chile

Marcelo Velasco
Ignacio Nieto

Acknowledgments

To write this book several people helped in different ways. We thank Karamjit Gill, Emeritus Professor at the University of Cambridge, for contacting us with Springer. To both Enrique Zamudio, Dean of the Faculty of Arts at the Universidad Finis Terrae, and Natalia Calderón, Professor at the School of Cinema at the Universidad de Valparaíso, we give thanks for creating opportunities for us to discuss and debate this topic. To Prof. Juan Carlos Letelier, Science Faculty University of Chile, for his always inspirative example and advice; to Ph.D. Juan Carlos Torres, for his invaluable help in translation; we are grateful to Santiago Aranguiz and Benito Escobar, Publication Coordinators in the Research and Publication Department at the Universidad Finis Terrae, for their comments on an initial draft of this publication. We also would like to thank visual artist and Professor Natascha de Cortillas from the Department of Arts at the Universidad de Concepción for her support; to journalist Camila Calderón for her help and commitment in this book; to the Government of Chile, Ministry of Culture, Arts and Heritage, National Funds for the Arts 2023 Research Line, fund N° 654589; and Nélida Pohl, Director of Communications at the Institute of Ecology and Biodiversity, for her critical interest in our research. Also the Art + Science Knowledge Building and Sharing in the XXI Century (ASKXXI) initiative for the opportunity to learn about art-science in the United States and the south of Chile; and María José Velasco, a last-minute hero.

We are grateful to many people and organisations for allowing us to use the images displayed in this book including Robert Pepperell, Artist and Professor of Cardiff School of Art and Design, as well as Ionat Zurr and Oron Catts, from SymbioticA, Centre of Excellence in Biological Arts; to Justyna Niewiara, Archivist of the Richard Deacon artwork; Dr. Luise Reitstätter, Cultural Scientist from CReA Laboratory for Cognitive Research in Art History at the University of Heidelberg of the Art; Zoya Dare, also from CReA; to Mariana Arellano at Fundación Engel; to Jungyeon Park from Studio Locus, South Korea; to Dave Furness, Professor Emeritus of Cellular Neuroscience, Keele University; to Ph.D. Anne Rothfeld, from the National Library of Medicine.

We would also like to give our gratitude to the sculptor and photographer Bogomir Ecker; Carrie Greenwood at Hagströmer Library from the Medical History and Heritage Karolinska Institutet in Stockholm; Tina Moo Young from HHMI and Sandy Geis for their generosity sharing Irving Geis illustrations; and Martin Hieslmair from the Corporate Communications/ Web Department associated to the Ars Electronica.

We also want to deeply thank the 22 interviewees, renowned artists and scientists from around the world, for providing us not only with contextual descriptions of their art-science work, but also technical information, theoretical insight and, crucially, permission to publish their images: Jean-Baptiste Boulé, Nicolas Desprat and Philippe Parreno; Dmitry Bulatov and Alexey Chebykin; Juan M. Castro and the Proto-Alien Project; Shanthi Chandrasekar; Suzanne Dikker, Matthias Oostrik and Suzan Tunca; Gilberto Esparza; David Goodsell; Kathy High; Yunchul Kim; François-Joseph Lapointe and Marianne Cloutier; Max Liboiron; Rachel Mayeri; Andro Montoya Riveros; Špela Petrič; Perdita Phillips; Jimena Royo-Letelier, Pierre Berger and Vincent Martial; Michail Rybakov; Susana Soares; Daina Taimina; Manuel Théry; and Paul Vanouse and Nicholas Wade.

Last but not least, all our gratitude goes to Gonzalo Cordova for his patience and confidence in this project, to Jayanthi Narayan and to Mimi Bick, for her enormous and generous work as Editor, Advisor and English Corrector.

About This Book

This book delves into the long-standing human aspiration to integrate art and science. In six chapters, *The Art-Science Symbiosis* outlines new approaches to understand current scientific practice in general and art-science in particular, showcasing how contemporary art can provide a unique perspective on the meaning and potential of interdisciplinary collaboration. With more than a hundred full colour images, *The Art-Science Symbiosis* may serve as a resource for those interested in researching on and applying the art-science integration, as well as a general reference for anyone interested in interdisciplinary and transdisciplinary work.

In the book, 22 artworks have been selected based on their inherent merits and for the emergent knowledge that their art-science integration produces. These works have sparked novel questions, ideas and curiosity among scientists and artists alike which, we hope, will promote further dialogue not only among them but with the general public, inspiring a process that may lead to diverse, complex and promising results with real-world consequences we have as yet to uncover. The key issues of the book are:

- Contemporary art is a powerful space of dialogue between science and the public.
- Interdisciplinary collaborations should be symmetrical and collaborative.
- Artistic inquiry can lead to new understanding of scientific exploration.
- Art-science has a strong transformation potential.
- The arts and the sciences are human expressions that anyone can explore.

Art-scientists contributing with their artworks

Jean-Baptiste Boulé, Nicolas Desprat and Philippe Parreno
Pierre Berger, Jimena Royo-Letelier and Vincent Martial
Dmitry Bulatov and Alexey Chebykin
Shanthi Chandrasekar
Suzanne Dikker, Matthias Oostrik, Suzan Tunca, and Than Van Nispen
Gilberto Esparza
David Goodsell
Kathy High
Yunchul Kim

François-Joseph Lapointe and Marianne Cloutier
Max Liboiron
Rachel Mayeri
Andro Montoya Riveros
Špela Petrič
Perdita Phillips
Juan M. Castro and the Proto-Alien Project
Michail Rybakov
Susana Soares
Daina Taimina
Manuel Théry
Paul Vanouse
Nicholas Wade

Contents

1 Introduction to the Art-Science Symbiosis 1
Bibliography 4
2 Challenges for the Sciences and the Arts 5
2.1 Challenges in the World of Science 6
2.2 New Ways of Understanding Science 9
2.3 Challenges in the World of the Arts 11
2.4 Why Integrate Art and Science? 14
2.5 A Point of Encounter: Both Art and Science Rely on Metaphors 15
Bibliography 17
3 The Diversity of Art-Science Integrations 21
3.1 Beginnings and Expansions of the Relationship Between the Arts and Sciences 21
3.2 An Analytical Framework for Understanding Art-Science Integration 27
3.3 Different Categories of Art-Science Manifestations 31
References 62
Bibliography 63
4 Selected Art-Science Works 65
4.1 Introduction 65
4.2 Art-Scientist, Selected Works 66
4.3 Art-Science Work Descriptions by Their Authors 67
4.3.1 *Immersion of the Henri Poincaré Institute in the R^3/Z^3 Space.* Pierre Berger, Vincent Martial and Jimena Royo-Letelier, 2017 67
4.3.2 *Evolvable Yeast Automaton.* Jean-Baptiste Boulé, Nicolas Desprat and Philippe Parreno, 2016 69
4.3.3 *That Which Lives in Me.* Dmitry Bulatov and Alexey Chebykin, 2011 72
4.3.4 *Formata. Non-terrestrial, Kinetic and Autonomous Agents.* PЯOTO-ALIEИ PЯOJECT, 2020–2023 75
4.3.5 *Kolam.* Shanthi Chandrasekar, 2021 78

4.3.6 *Harmonic Dissonance: Brains and Bodies Dancing Together at the Interface of Art and Science.* Matthias Oostrik, Suzan Tunca, Than Van Nispen, and Suzanne Dikker, 2018–2023 (Ongoing) 81
4.3.7 *Nomadic Plant*, Gilberto Esparza. 2008–2013. 85
4.3.8 *Integrative Illustration of Cells and Viruses.* David Goodsell, 2020 88
4.3.9 *Blood Wars.* Kathy High, 2010–2011 89
4.3.10 *Impulse.* Yunchul Kim, 2018 94
4.3.11 *Microbiome Rebirth Incubator.* François-Joseph Lapointe and Marianne Cloutier, 2016–2023 97
4.3.12 *Plastic Is Land.* Max Liboiron, 2018. 100
4.3.13 Primate Cinema: Apes as Family. Rachel Mayeri, 2011 103
4.3.14 *Creatures.* Andro Montoya Riveros, 2008. . . 106
4.3.15 *Solar Displacement.* Špela Petrič, 2014 110
4.3.16 *Green, Grey or Dull Silver.* Perdita Phillips, 2007–2008 113
4.3.17 *Inhabiting a Map.* Michail Rybakov, 2018. . . 117
4.3.18 *Bee's.* Susana Soares, 2007–2009 121
4.3.19 *Crocheted Hyperbolic Planes.* Daina Taimina, 2009–2014 124
4.3.20 *Living Architectures*, Manuel Théry. 2008, 2013 127
4.3.21 *Labor.* Paul Vanouse, 2019. 129
4.3.22 *Neuroportraits.* Nicholas Wade, 2011–2012 133
4.4 Biographies. 136
5 Capturing the Insights Behind the Selected Art-Science Works 145
5.1 Motivation to Engage in Art-Science 146
5.2 Scientific Topics and Lines of Research. 148
5.3 Materiality and Artistic Methodology 150
5.4 Differences Between Artistic and Scientific Approaches. 151
5.5 Art-Science Synthesis: Its Meaning and Implications 155
6 Action Points and Perspectives 159
6.1 Work on Your Own Art-Science Project. 160
6.2 Current Art-Science Initiatives. 161
6.3 Postgraduate Programs. 166
Bibliography. 167

About the Authors

Marcelo Velasco (Santiago, 1971) is a biologist with graduate studies in Medial Arts (Universidad de Chile) and Ecological Economics (Universitat Autònoma de Barcelona). As an experimental biologist, he worked in Neurosciences studying the epistemology, physiology and behaviour of the visual system and magnetic perception. During the 2000s, he developed studies for the sustainability performance of various industries. He has developed and applied innovative social instruments for collaboration and inclusion using new technologies and interdisciplinary approaches. This and other works resulted in presentations in Argentina, Colombia, Canada, the UK and throughout Chile. Currently, he is devoted to unfolding the transformative potential of art-science inside and outside the scientific community.

Ignacio Nieto (Santiago, 1973) studied Fine Arts, has a Masters in Educational Informatics and is currently a Ph.D. candidate in Philosophy with a specialisation in Aesthetics and Art Theory at the Universidad de Chile. For the last 15 years he has worked in several Chilean universities, teaching object-oriented programming to design and art students, making expositions, and as a curator, giving lectures and workshops on new media. His proposals have been presented in different places in Chile and in different countries: Argentina, Brazil, Colombia, Mexico, USA, Switzerland and the Netherlands. He has also been a juror for the Apex Art Gallery in New York, the Symposium for Electronic Arts, ISEA, and guest editor of the journal AI and Society. He is currently completing his doctoral thesis, based on the representations that take into account, in the art practice.

1 Introduction to the Art-Science Symbiosis

Abstract

This introductory Chapter depicts the authors' motivations in writing this book, describes its structure and briefly outlines the objectives, premises, borders and expectations assumed in this interdisciplinary endeavour. Some basic qualitative differences that are found in works of art-science integration or symbiosis are identified, as are the basic assumptions underlying the approach chosen by the authors.

Keywords

Art-science · Art-scientists · Interdisciplinarity · Science communication

This book delves into the long-standing human aspiration to integrate art and science and proposes new possibilities for their harmonious coexistence. Through a review of art-science works, it showcases how contemporary art can provide a unique perspective on cutting-edge scientific research, exploring the intersection of art and science across different historical periods, contexts, and purposes, highlighting the meaning and potential of interdisciplinary collaboration. We have been trained and are immersed in both fields, with backgrounds in biology, neuroscience, medial arts, visual arts, programming, education, and philosophy. We challenge each other to approach the subject from diverse perspectives and share our enthusiasm for exploring, deepening, and pushing the boundaries of this integration.

The goal of *Art-Science Symbiosis* is to serve as a resource for researchers interested in interdisciplinary and transdisciplinary work that bridges the fields of art and science, as well as a general reference for anyone interested in the topic. Due to the complexity and breadth of artistic expressions, this publication focuses solely on the visual arts, involving the production and exhibition of objects created using various techniques.

The study of art-science integration through analytical approaches has received little academic exploration. However, there is a significant body of literature, exhibitions, and biennials that focus on the relationship between art and technology, which is often confused with the art-science relationship. To differentiate between the two, we may understand the art-science relationship here closer to "art and basic science", where technology is a means of advancing knowledge without treating art as just a tool for applied scientific procedures.

As students and then practitioners of art, we have noticed that many artists are interested in scientific topics and techniques. These interests

© The Author(s), under exclusive license to Springer Nature Switzerland AG 2024
M. Velasco and I. Nieto, *The Art-Science Symbiosis*, https://doi.org/10.1007/978-3-031-47404-0_1

often influence the development and proposals of their artistic works. We have also observed that scientists are often unfamiliar with the cultural and philosophical implications associated with the premises of contemporary art and with questions surrounding positivism (the so-called "crisis of modernity"), an important topic of reflection in the twenty-first century. These observations led us to wonder whether art too can have an impact on science and if so, whether it can make a valuable contribution to scientific inquiry.

In general, art-science works stand out for their artistic merit, yet contributing little to the scientific field that informs them. In other cases, the work harnesses artistic skills solely for scientific purposes. In both cases, either science or art held greater authority, defining the scope of the integration and creating a hierarchical relationship that served the disciplinary purposes of the dominant field. Our training in art and science allowed us to bring a double sensitivity to what we encountered. Thus, the most strong inspiration came with artworks that captured our attention for both artistic and scientific merits, but also by the emergent knowledge that the art-science integration produced. These works existed in an intriguing intermediate zone, where the authority of art and science was notably symmetrical and their mutual influence visible.

Initially, the search focused on symmetrical works created by professional artists who incorporated science into their practice. However, our search yielded some fortunate findings, leading us to expand its scope to include works created by professional scientists integrating contemporary art into their scientific practice. When scientists integrate the arts and the sciences in a symmetrical way, they expose themselves to new audiences, new questions and critical viewpoints, placing them on a more horizontal plane in relation to the public, as discussed by Born and Barry (2010). They spark new questions, ideas and curiosity among scientists, artists, and the public, facilitating the crucial dialogue that addresses the complex relationship between subjective perspectives, emotional experiences and objectivity. In this approach the subjective choices made by scientists in the selection of research questions or forms of capturing evidence are made more visible.

For artists, a symmetrical integration allows for expanded objectual possibilities and meanings, leading to complex visual devices. Experimental science forces artists to consider empirical constraints to their projects, for example, in terms of the replicability or the explanatory power of the proposal. By breaking down the barriers between the sciences and arts, we may discover new and enigmatic paths that lead to diverse, complex, and promising results with real-world consequences that we have yet to uncover.

In choosing works for this book in a process akin to a curatorship in the world of art,[1] we resort to various conceptual criteria, including: relevant integration; mutual transformation; scientific disruption through art; the poetics of science; art and objectivity. We have selected 22 splendid works that embody a virtuous symbiosis of art and science, which are presented in Chap. 4. All of them challenge both art and science while addressing specific issues like learning, identity, excessive material growth, or our coexistence with the natural and the artificial.

This selection process raised many further complex questions: Is it possible to propose criteria that links or values these works as an artistic movement? Can we assess the contribution to science of these proposals and research perspectives? What specific role does art and science play in each work and ultimately what is the epistemological status of these hybrids? How and what can we learn from the knowledge that emerges from these interdisciplinary operations and mutual transformations? It is important to warn the reader that this book does not aim to provide thorough answers to these questions. Rather, it seeks to make them explicit by linking them to functional theoretical proposals which advance an open and timely dialogue between

[1] *Curatorship* is the process of making "decisions about which works of art should be displayed and how they should be seen. They also decide on the themes and ideas to be experienced by the audience." (AGSA 2023, parr 4).

art and science, an invitation we extend to the readers.

Chapter 2 expounds a critical approach towards both art and science that describes some of their disciplinary problems that may advocate on the relevance of a transformative effort through interdisciplinary practices. To observe this integrative phenomenon more widely, Chap. 3 provides a retrospective of the constellation of art-science manifestations presenting references and examples from different places and historical periods. Based on this review, we propose a classification of six types of works, which employs as its underlying elements the relation of authority between art and science; the scope of influence of the results obtained; the way in which the results are presented or exhibited; and the vocation for mutual transformation that emerges from the knowledge the work generates In relation to the images selected for this chapter, we resort first to the high quality free access versions of them.

Chapter 4 presents the selected 22 current works which are displayed in alphabetical order according to the surname of one of their authors. Each work is described by its creators and is accompanied by images generously provided by them. These art-scientists work in different parts of the world, with many hailing from "remote" places such as Santiago, Chile; Tallinn, Estonia; Bogota, Colombia; Chennai, India; Perth, Australia; or Kaliningrad, Russian Federation. The selection includes both artists who draw inspiration from science and scientists who venture into the realm of contemporary art, whose artworks, valuable in both artistic and scientific terms, define the basis of the curatorial process. These works invite us to reconsider the processes and outcomes of both science and art, opening a broader discussion on the challenging task of knowledge generation.

In Chap. 5, we present an analysis of the semi-structured online interviews we conducted with the art-scientists presented in Chap. 4, based on a predefined questionnaire (the whole team involved in each work was not always present in the interview). The aim of these interviews is to provide testimonies on how these hybrid processes are embodied, understood, and projected by the authors. It is from this perspective that we examine and contrast the methods, motivations and strategies of art-science integration and hope it will contribute to future analytical frameworks and the development of an art-science metalanguage.

Finally, Chap. 6 invites the reader to get involved in an art-science project. Some easy to follow steps are presented as a guide to transform a scientific research passion into an art-science proposal. Several art-science current initiatives are also presented with a list of salient art-science graduate programs around the world.

The book was written entirely in the city of Santiago, Chile over the course of four years. Far from major contemporary art centres we learned about various institutional efforts to link the work of artists and scientists in a virtuous manner.[2] Some of these consist of relatively established efforts, such as Media Labs, which are spaces for cooperative work and learning focused on the use of new media and technology. Pioneering graduate programs that provide interdisciplinary training in art and science have also emerged.

Some art-science linkage efforts have been described in articles, books and art catalogues. In some of these initiatives, artists are invited, without a very structured plan, to "reside" in a laboratory or scientific research centre for a certain period of time, so as to generate a piece of work inspired by the observations and conversations that occur there. Other initiatives foster conversations between the artist and the scientist during the production process of the work itself. Unfortunately, the results of many of these

[2] An inspiring work developed in Chile in the late 60s, was the pioneer public performance displayed by Chilean physicists Carlos Martinoya and Nahum Joël. Using a device called the *Chromatic Abstractoscope: The Robot of Abstract Painting* (1969), and the physics of polarised light and the singular properties of crystals, generated images comparable to the works of abstract painters. The demonstration was presented in an open-air art exhibition along the river Mapocho in Santiago. "Some artists felt that we were only demonstrating another undesirable intrusion of science into art. Most of them, however, were impressed and saw… a powerful source of artistic inspiration" (Martinoya & Nahum 1968, p. 173).

efforts are uninteresting to art and irrelevant to science. In our view, the process fails to overcome the asymmetries inherent in the encounter between the illiterate and the expert, making relevant integration difficult. Put otherwise, a spontaneous dialogue between two very distinct profiles -artist and scientist- does not happen easily and when it does it often ends up being confusing, exhausting, and frustrating for both.

From an *anti-disciplinary* standpoint,[3] we seek to explore this methodological gap in order to stimulate more productive dialogues that integrate art and science based on two premises: (1) that there need be an explicit discussion between artists and scientists regarding their expectations as to who will have the authority to determine the purposes and processes of the project; and (2) that there need be a mediation of the interaction between scientists and artists. We discovered that artists and scientists require adequate "translations" of their respective disciplinary languages to understand the others' goals, emphases, ambitions and prejudices about the aforementioned hierarchy of authority. Likewise, to become acquainted with the history of the "other" domain, and to understand the common day-to-day life of its practitioners along with their difficulties and worldviews. We saw that the mediation and accompaniment through an agent who can move between the two worlds is useful for a fruitful dialogue in the eyes of both artists and scientists. In many cases, as in several of the selected works presented in Chap. 4, this agent is embodied in the creators of the work themselves when they have the privilege of having a solid background or practice in both art and science.

Some clarifications about the terminology used in this book might be of use to the reader. We use the compound term "art-science" to refer both to the integration of art with science, and science with art and, by extension, we refer to its authors as "art-scientists", irrespective of whether they consider themselves scientists or artists in their professional life. The title of this book employs the concept of symbiosis, which is a metaphor for the intimate association or close union of two different "organisms", in this case, the arts and the sciences. This association can range from circumstantial encounters that may be relevant to only one of the organisms involved, to a mutual transformation where both organisms benefit or even survive from the association. To be more precise (and less metaphorical) throughout the text we mostly use the term "art-science integration" to refer to any type of interaction between science and art practice, or between art and science practice.

Lastly, this book does not view art as a means of generating knowledge in the same way as traditional experimental science. Instead, it explores how artistic methodologies intersect with experimental science to produce unique results. By materialising speculative and critical proposals of science through art, the book showcases experiments that are not always replicable and proposes ideas to both non-specialist audiences and peers within a new context.

[3]As expressed by Joichi Ito, the concept of the antidisciplinar "it's about working in spaces that simply do not fit into any existing academic discipline—a specific field of study with its own particular words, frameworks, and methods." (Ito 2017, p. 4). "With funding and people focused on the disciplines, it takes more and more effort and resources to make a unique contribution. While the space between and beyond the disciplines can be academically risky, it often has less competition; requires fewer resources to try promising, unorthodox approaches; and provides the potential to have tremendous impact by unlocking connections between existing disciplines that are not well connected." (p. 5).

Bibliography

AGSA. (2023). *What is a curator?* Art Gallery of South Australia. https://www.agsa.sa.gov.au/education/resources-educators/agsa-art-school-online/what-curator/. Reviewed June 12, 2023.

Born G and Barry A (2010). ART-SCIENCE, From public understanding to public experiment. *Journal of Cultural Economy*, 3:1, 103–119.

Ito J (2017). Design and science. *Journal of Design and Science*. https://jods.mitpress.mit.edu/pub/designand-science/release/2. Reviewed August 2023.

Martinoya C and Nahum C (1968). The 'Chromatic Abstractoscope': An Application of Polarized Light. *Leonardo*. Volume 1, pages 171–173.

Challenges for the Sciences and the Arts

2

Abstract

This chapter elaborates on some of the challenges currently faced by science and art, encouraging a greater interdisciplinary openness to address them. New approaches to understanding scientific practice and the world of art are reviewed. Finally, the chapter discusses ways that science might be made more visible in non-scientific contexts, and how art and science can be enriched by other disciplines, using non-conventional communication channels.

Keywords

Art criticism · Sciences · Experimental system · Scientific pluralism · Artistic research · Interdisciplinary research

The world is facing a crisis of truthfulness. Many people believe in science but feel disconnected from it. They view scientists as isolated in their ivory towers, disengaged from real-world circumstances. Simplistic dichotomies appear and take hold, such as basic science versus applied science, or science by curiosity versus science by mandate. Gibbons (1999) argues that a new social contract is needed to re-establish the relation between science and the public if science is to be empowered and validated. Despite the emergence of currents of thought such as postmodernity and deconstruction, the culture of the hard sciences has not yet fully embraced any epistemological shift (Dupuy & Varela 1995).

The field of science confronts many challenges. One of these is the crisis of meaning of scientific careers, as students feel isolated from the real-world problems their communities face (Saltelli & Funtowicz 2017). Moreover, new perspectives question the positivistic stance of science, in particular its claim to strict objectivity. Haraway (2019) argues that objective science cannot be separated from the subjective narrative that underpins it. The interdependence of objectivity and subjectivity is complex and conflicted, however political or gender-related assumptions are often hidden beneath the veneer of objectivity. Meillassoux (2020) suggests that scientific statements should be understood as a sign of the times which reveal our cultures' systems of beliefs. Is it possible for scientific practice to step outside its hyper-specialisation to reflect critically on the context in which it operates?

On the other hand, the contemporary art world has become increasingly commercialised, with many artistic operations oriented towards the sale of a product. Like science, art has become self-referential, lacking the capacity to criticise or formulate new representations that offer relevant, informed, and reflective commentary regarding the complex world we

© The Author(s), under exclusive license to Springer Nature Switzerland AG 2024

M. Velasco and I. Nieto, *The Art-Science Symbiosis*, https://doi.org/10.1007/978-3-031-47404-0_2

inhabit. Art has become distant from the world, relying on vague conceptual ideas and exhausting its material, symbolic and representational resources. In addition, the art world's self-referentiality has led to repetitive "quoting" of other artists, perpetuating a cycle of reiteration that is valued primarily by peers seeking acceptance in the world of art. These traits have contributed to a sense of crisis (Rojas 2012), resulting in the diminishment of art's critical validity and its potential to disrupt and dissent. The disruptive and transgressive aspects of art have been co-opted by their market appeal thereby infusing artistic representations with a degree of cynicism. As a speculative commodity, contemporary artworks are managed by trendy galleries and elite groups. According to the Artprice database that tracks these sales, in 2022 the average price of works by artists under the age of 35 was nearly $28,000 (Azimi 2022). Meanwhile, talented artists living outside the bounds of this elite system work in precarious conditions, particularly in the third world.

This chapter elaborates on some of the challenges faced by science and art, in the hope of encouraging a greater interdisciplinary openness to address them. Beginning with the realm of science, the topics of job dissatisfaction and unhealthy workloads, the replicability crisis, power asymmetries in the lab, disconnection from real-world problems, and the dilemma regarding articles never cited will be briefly portrayed. Then, alternative approaches to understand the scientific practice are presented, as a conceptual openness for interdisciplinary efforts. Looking at the world of art, the discussion revolves through the relation between academia and broader artistic practice, the questions surrounding art circulation and finance, and the debate about the meaning and relevance of artwork today. A discussion is presented on the ways science might be made more visible in non-scientific contexts, how it can be enriched by other disciplines and transmitted using non-conventional communication channels. Finally, the use of metaphor is proposed as a powerful point of encounter between the arts and the sciences.

2.1 Challenges in the World of Science

Despite the inherent satisfactions of exploration and discovery, being an academic scientist today poses some very specific challenges, particularly for those who are at the start of their careers (Nature Editorial 2019a). Competition in the field is fierce as is the pressure to produce and publish conclusive results. In industries that hire scientists, intellectual abilities are usually recognized and better remunerated than in academia. According to *Nature*, scientists' job satisfaction is at an all-time low, with less than 60% of the 3200 scientists recently surveyed satisfied with job security and the possibility of developing their projects with complete freedom (Woolston 2021).

An important mark of scientific activity is replicability: experiments should be reproducible in different laboratories by different scientists; given identical experimental conditions, the same results should consistently be obtained. According to several authors, many scientific studies suffer from a replicability crisis, particularly in the areas of experimental psychology and behavioural sciences, but also in biology and other hard sciences (Serra-Garcia and Gneezy 2021; Clark 2021; Lehrer 2010; Haelle 2021; Gosselin 2020). Nevertheless, some experimental conditions may be so expensive or require such a sophisticated and often scarce level of craft and expertise, that reproducibility becomes almost impossible. If a new experimental approach does not arrive at the same results as the original experiment, this does not mean, however, that the results of the original experiment were wrong or flawed. The new setup may be "seeing" or "not seeing" the same aspects of the problem. Some authors, like Bernard, argue that mathematics is the only field where reproducibility must always be expected, since its arguments are based on axiomatic logic (2023). But the experimental sciences, may be argued, do not resort to such axioms, and they should not be held up to the bar of strict replicability.

There is a budding belief that another reason for the lack of reproducibility in scientific experiments is faulty communication, particularly in attempts to replicate complex experiments which have been described in unclear or obscure terms (García-Bullé 2021). To address this issue, a movement has emerged advocating reforms to promote more open and accessible science through, for example, the prior registration of experimental results, incentives for more rigorous research, and the implementation of a controlled and auditable system (Munafò et al. 2017).

Another challenge facing science are asymmetries of power within its structures and practices. These can be usefully illustrated by considering the history of the Nobel prize. Since it was first awarded in 1901, only four women have won the Nobel in Physics, twelve in Physiology or Medicine, and seven in Chemistry. Nobel prizes have been repeatedly awarded to the same demographic group representing the flaws in the recognition and rewards in science (Sanderson 2021). Amanda Moehring, a behavioural geneticist at Western University in Canada, shows that the proportion of female physicists earning PhDs since the 1980s far exceeds their representation among Nobel laureates in physics. "The fact that white men are so radically overrepresented among Nobel laureates in science tells us in no uncertain terms that we, as a society, are missing out on a wide variety of new ideas and discoveries that depend on a diversity of ideas and backgrounds" (op. cit. parr 5).

Asymmetries of power are not only a matter of gender. As in other fields, personal characteristics such as age, ethnicity and place of birth, as well as the stated purpose behind research all intervene in scientific practice, affecting the scope and diversity of scientific knowledge generation.

When the primary source of information for public decision-making is science, it is essential that scientific practice be marked by confidence and responsibility. Although at least in the United States public trust in scientists is on the rise (Ledford 2019), there is still a notorious lack of scientific input in decision-making, mainly due to a persistent sense of mistrust between scientists and politicians (Nature editorial 2019b). This is particularly true in relation to crucial socio-environmental issues of our time, such as biodiversity and climate change. Arguably scientists need to find different ways to make their voices heard and overcome the paradox that by pretending to be outside and above politics, science has in fact become politicised (Howe 2020). Another pressing issue is the disconnection between scientists and real-world issues, as well as the public's disengagement from science given the complexity and specialisation of scientific fields and limited understanding of how science works as a system.

Since the 1800s, the explosive development of science has led to the growth of increasingly specialised sub-disciplines which continue to emerge. While it was commonplace in the nineteenth century for a scientist to be a geologist, a botanist and a zoologist, nowadays a zoologist might easily struggle to understand their colleague who works in a very different area or speciality. Similarly, in the past a person with a basic education could understand quite a bit about science if he or she was interested in it. The same cannot be said today. Knowledge of the processes of science among those who do not have any specific scientific training is extremely scarce. This "black box" of science hardly contributes to the generation of trust and participation. Along these lines, an important challenge for the relationship between science and the public is the inaccessible character of highly specialised peer-reviewed papers which detail the procedures, results and conclusions of research, excluding all factors that do not directly and immediately impact the object of study. They are condensed forms of writing that refer to complex subjects which may involve several authors (sometimes even hundreds). The language used is highly technical and therefore inappropriate for general audiences. "Academic writing is not something we can all understand, it is a code that only members of an elite can decipher" (Becker 2011, p. 51, translated by the authors).

Social support for scientific research is partly based on a social contract founded on what Ladd et al. call the "myth of accountability" (2009 p. 2), that is the idea that internal scientific integrity is tantamount or equal to social responsibility. According to these authors, one way to genuinely improve social control over scientific achievement is to increase the emphasis on social considerations in the financial support and evaluation of research projects, beginning with the training of scientists to be more mindful of and explicit about the social implications of their work. In this sense, life science researchers may lack awareness of the relevance of ethical and societal concerns to their research, and when they are aware, they often lack the self-confidence to broach them openly (McCormick et al. 2012). Stroffolino (2019) concludes that scientists in general need to be trained more rigorously in effective interaction with the public, to deal with ethical issues and the communication of risks associated with their research.

Additionally, the peer review process, which is an essential component of the scientific enterprise, could be more robust by adding criteria which assess research based on its broader societal impact. It is suggested that "open access to the record of science is essential for an equitable and inclusive global scientific enterprise and to the scientific self-correction that is crucial for rigour and public trust" (Boulton et al. 2021, parr. 1). This is the conclusion of a correspondence published in Nature regarding the statement made by the International Science Council (ISC) and All European Academies (ALLEA) about the need to reform open access to scientific records (op. cit). It is worth nothing that governments not only subsidise research work, but also support the publishing industry, which relies heavily on scientific content.

Today, the relevance of scientific research is often measured by the number of citations a publication accumulates. Citations may represent the influence of the investigative work on the generation of knowledge in the subject and even beyond it. Nonetheless, there is an alarming lack of citations in many research fields and some investigations do not exhibit any impact at all (if measured by the quantity of citations). Of 39 million research articles across all disciplines recorded in the Web of Science between 1900 and 2015, roughly 21% have never been cited. Most of these uncited articles appear in little-known journals; conversely, almost all articles published in well-known journals have been cited (Van Noorden 2017; Eveleth 2014). Paraphrasing Poincaré, much of science is devoted to making bricks but few scientists erect buildings.[1]

In terms of its disruptive capacity, that is, those scientific contributions with the potential to change how a problem is approached, a recent study concluded that science is more self-referential and less disruptive compared to the past (Park et al. 2023). While research in the 1950s tended to use words that evoke creation and discovery such as "produce" or "determine", research in the 2010s referred most frequently to incremental progress, with terms like "improve" or "enhance" appearing most often (Kozlov 2023). Neuroscientist Elaine Sevier is clear in this respect:

> Everyone wants to write the classic paper that is still referenced in 100 years. But to advance your career, you need an easily digestible story that will be immediately popular—and you should seek, interpret, and present your data accordingly. This is a constant cycle of short-term thinking that does not tolerate slow, sometimes boring, sometimes weird work that lies outside the mainstream. That also means it does not tolerate the work that has the potential to become disruptive. (2023, parr. 8)

The high degree of specialisation required to master a scientific field often means that scientists must focus their research efforts on a narrow set of experimental resources. While this can lead to advances in the field, it can also result in research that is isolated and limited in

[1] This sentence comes from Poincaré: "Science is built of facts the way a house is built of bricks: but an accumulation of facts is no more science than a pile of bricks is a house." Oxford Reference (2023) https://www.oxfordreference.com/display/10.1093/acref/9780191866692.001.0001/q-oro-ed6-00008396. For an explanation of this comparison see Forscher B (1963) Chaos in the Brickyard. *Science*. Vol. 142, Issue 3590 p. 339 https://doi.org/10.1126/science.142.3590.339.a.

scope. This narrow focus can, again, lead to a recursive and endogamic approach to scientific inquiry, with researchers becoming less willing to take risks and explore new avenues of research. In the light of these challenges, some thorny questions about the value of science and who ultimately benefits from it necessarily arise. Confidence also owes to factors such as the degree to which citizens trust state institutions more generally, as well as the level of inequality in a society (Gallup 2019). Nevertheless, the successful development of drugs, vaccines and other medical interventions to combat the Covid-19 pandemic have shown the promise of science to contribute to building healthier and, ideally, more resilient, equitable and just societies.

2.2 New Ways of Understanding Science

The challenges already mentioned prompt a re-examination of traditional scientific practice. One of these concerns the conditions that need to be established to validate scientific knowledge. Advocates of these demands argue for a more nuanced approach that considers relevant contextual variables and recognises the limitations of reductionist results. In addition to these proposals, the fields of history, anthropology and philosophy have generated compelling arguments questioning some of the premises associated with experimental work. This has led to the development of a new understanding of the scientific production process, and a space for interdisciplinary dialogue that may ultimately result in more robust and innovative knowledge proposals.

In this sense, current lines of thought open science to other fields of inquiry by suggesting that complex phenomena can be better explained from multiple perspectives or disciplines rather than a single one (Kellert et al. 2006). There are several "pluralist" positions in the philosophy of science all of which consider the aspiration to unify the sciences as faulty, pretentious or reductionist (Ludwig and Ruphy 2021). According to scientific pluralism, sometimes known as the Stanford School (also labelled the "Stanford Disunity Mafia"), neither the languages of scientific disciplines nor their subject matter is reducible to one language and one subject (Suppes 1978). Cartwright (1999) argues that science is a patchwork of laws with no single one that is fundamental to them all. Longino (1990) describes aspects of the socio-cultural activity in which scientific knowledge is structured and Latour and Woolgar (1995) invites us to reconceptualize science in terms of a complex relationship between practices, instruments and institutions.

Many factors that are extrinsic to the scientific method have been distinguished as influencing scientific results. Perhaps the most relevant of these is the context which accounts for the questions or objectives addressed (or ignored) by scientists and scientific institutions, and endorses values or trends of a particular economic or political nature (Kellert et al. 2006). This subject is infrequently discussed by scientists themselves but is of keen interest and concern to philosophers and sociologists of science.

Along with other authors –Knorr Cetina, John Law, Annemarie Mol, Madeleine Akrich and Michael Callon–Bruno Latour analyses scientific practice from an ethnographic perspective and examines it in terms of *networks of power relations between actors* (Latour 2007). Latour's widely debated position eliminates the distinction between the natural and the social and posits scientific practice as a space of encounter between the human sciences, nature, and technology. This encounter includes humans, such as scientists, public servants and financiers, non-human actors or materials, such as computers, instruments and machines, and immaterial objects like algorithms, methods and taboos. Participants in these networks of power, collectively referred to as *actants*, make and break alliances amongst themselves and with respect to other networks, around what may or may not be "true". This is how ideas are adopted or rejected, generating a space of relationships in permanent co-production. From this perspective, the central question is which actants construct scientific claims in a specific cultural setting (Latour 1992).

This way of understanding the scientific process upends the seventeenth century approach to scientific knowledge, by which correctly posed questions, well-constructed instruments and carefully conducted experiments give rise to immutable scientific conclusions. Contrarily, Latour claims that entities like viruses or galaxies are never experienced in a direct or univocal manner, but are conceptually stabilised through "…different wavelengths, different encodings, different image treatments. The phenomenon resides not so much in an image in itself as in the superposition of a multitude of different indications … which are assumed constant through the transformations of graphs, tables, photographs, reports, calculations" (Latour 2010, p. 137). For Bachelard (1984), scientific instruments can be considered "materialised theories", implying that contemporary science "thinks" through its apparatus. Additionally, with modern means of recording and exchange, the dispersion, singularity or scarcity of evidence can be avoided or diminished by capturing, collecting and comparing available evidence. With systematised evidence at hand, anecdotal dissent becomes increasingly difficult to sustain. This process commonly culminates in the publication of a scientific article in a peer-reviewed journal which consolidates the evidence and discusses its implications. If successfully accepted, the idea gains allies and, therefore, power.

The purpose of these ideas is not to weaken or dilute the sciences with sociocultural relativism constructed on the basis of subjectivities, geographical, temporal and political contexts, and ethnic and gender biases, amongst other factors. On the contrary, it's purpose is "to explain the enormous consequences of science" and to understand "the mechanisms that make them possible and legitimate" (Nieto Olarte n.d., p. 4). Understanding the limits of science may foster and improve confidence in the method. Efforts in this direction have been made by Sormani et al. (2019) as well as Scott (2006). A one-way dialogue about what science is and what it should be falsely assumes that only scientists can talk about science, as if only economists could talk about economic problems, or ecologists about the environmental crisis. Interestingly enough, Latour presented, discussed and illustrated these ideas in an art space at the Zentrum für Kunst und Medien (Center for Art and Media Karlsruhe) in 2016, in a book and exhibition called *Reset Modernity.*

Other authors concur with this relational perspective. Following Thomas Kuhn, Hans-Jörg Rheinberger (1998) defines science in terms of discontinuity and relativity, epistemic practices or *forms* of approaching knowledge. Rheinberger develops an instructive terminology, specifically when analysing experimental practice in molecular biology. He describes the "experimental situation" (p. 287) as simultaneously local, social, institutional, technical and instrumental. In his view it is necessary to move from the idea of the single experiment to that of a system, since it is this system–with its institutions, devices, human and non-human agents, time and space–that provides the context in which the experiment makes sense (Schwab 2013). It is this situation which gives rise to the scientist's working unit, the *experimental system*, that harbours the *experimental object.* This system must be able to "differentially reproduce itself" and be subverted while retaining its reproductive coherence, as a "generator of surprises" (op. cit p. 287). Reproductive coherence is possible through the manipulation of a collection of stable *technical objects* such as instruments, infrastructure, algorithms, sensors and chemical substances which configure and stabilise the experimental system that, unlike the *experimental object* under manipulation, operates under control.

The stable technical framework –the setup– allows the experimental system to be systematically repeated by the researchers, altering its technical variables and components, and recording the effects of these alterations. In this way, the spaces of freedom of the experimental object are controlled and challenged in one or another dimension in a negotiation process with the scientist (Latour 2004) to find factors of stability or coherence relevant to the context of the experiment. This stability is made visible and takes shape through material traces: stains, fluorescent or radioactive marks, electromagnetic signals, absorbances. These traces are obtained through the technological objects as a reference of a stability robust enough to "control" the relevant aspect of the object. In this situation, the experimental object can become a technical object that may be part of a new experimental system. In this sense, the experimental object is open with respect to its future technical potential. Many technical objects used in daily lives are "packaged" with the technical conditions of the experimental system that brought them into existence in order to keep them stable and functional. Rheinberger (2021) draws on Jacob to argue that the experimental system is a "future-making machine" that does not just deliver answers but also shapes the question to be answered. It is a "device for materialising questions" (p. 288).

If there are so many variables with important roles to play in the construction of knowledge, might the arts be another relevant one? Schwab (2013) raises this question and wonders whether Rheinberger and Latour's ideas give way to the possibility of an "artistic mode of inquiry" (Rheinberger 2021, p. 288). Schwab explores possible links between artistic and scientific experimentation, comparing the laboratory to the artist's studio. Focused on material, temporal and experiential factors, he seeks a common research horizon for the natural phenomena of science and the cultural phenomena of art.

2.3 Challenges in the World of the Arts

James Acord is an artist who lived for fifteen years in the dormitory town of the Hanford nuclear production complex in Richland, Washington, USA. He explored and used the equations of nuclear physics to create his sculptural work. Accord was the only private individual in the world licensed to possess and handle radioactive materials. He acquired nuclear fuel rods containing depleted uranium from the terminated and inactive German SNR-300 reactor, and used it as material to make art. His licence number was tattooed on his neck. This case of commitment to both a scientific and artistic practice was an isolated case for many years. The knowledge acquired, in this case from physics, was included in the process and the artistic piece he created, generating provocative results (Acord 2009).

This example is both complex and unique, highlighting that in contemporary art, the traditional criteria for evaluation or interpretation are no longer applicable. As art broke away from traditional models with the emergence of artistic vanguards in the nineteenth and twentieth centuries, its appreciation has become vastly more complex, and boundaries have become blurred, leading to more questions than answers. There is no singular vision regarding the function of contemporary art nor precisely how to describe it. The formats used are also unconventional, with artists resorting to almost any possible materiality, size or time frame; the body, the gesture, or an immaterial concept; mixing and mashing of previous works; or the convergence of almost any former analog media in digital format. In this sense, contemporary art seems to exist in a grey area akin to a paradigm shift. Traditional criteria for the appreciation of art, such as harmony or fidelity to reality, have been replaced by a range of open interpretations, including contextual, diffuse, contradictory, emancipatory, disruptive, ethical, or even anti-ethical ones.

Schaeffer (2005, p.15) proposes that aesthetics or the appreciation of art has three dimensions. The first relates to judgement: a person can judge a work of art in different ways at different moments. The second centres on the relations between the parts: each part of the artwork can relate to another in different ways, and this can change the meaning of the work. Thirdly, judgement is deeply influenced by the artistic context, for example, the space where the artwork is exhibited or by the encounters that take place between artists, peers and critics during the creative process.

From a sociological point of view, Howard Becker introduced the term *Art Worlds* to describe "the network of people whose cooperative activity, organised via their joint knowledge of conventional means of doing things, produces the kind of art works the art world is noted for" (Becker 1982, p. x). This network includes artists, gallery owners, critics, promoters, general and specialised publications and other media, academies and academics, mentors, audiences, buyers and sellers. Artworks, then, are "not the products of individual makers, 'artists', who possess a rare and special gift. They are, rather, joint products of all the people who cooperate via an art world's characteristic conventions to bring works like that into existence" (p. 35). Art is "what an art world ratifies as art" (p. 156). The circular nature of the definition is apparent.

During the twentieth century, the relationship between art and technology has been the subject of an extensive bibliography, and of many exhibitions and biennials that treat its aesthetic, creative, material and strictly technological elements. Three milestones linked to representation and the use of technical elements stand out for their impact on changing the character of art. The first milestone is the avant-garde movement starting at the end of the nineteenth century with movements such as impressionism, expressionism and many other *-isms*. Here, art explores speculation and criticism, centred on the social impact of technology and industrialization. Such is the case of Italian futurism or Dadaism in the early 1900s, which abandons the static, integrates the machine, introduces fragmentation and represents time in sculpture, painting, or even on individual or social bodies, all of which were deeply influenced by hopelessness as a result of the world wars.

A second milestone is the appearance of cybernetics in art of the late 1940s. It critically revises the idea of representation and, based on a model of sensory response, adopts the view that machines can "observe" spectators.[2] Cybernetics also influences the counterculture of the 1970s through activities with computer research, media, and artistic experimentation with multiple exponents. A remarkable example of the link between technology, information and cybernetics in the arts was put into practice by the artist and kinetic sculptor Theo Jansen who, through computer modelling and the physical construction of mobile sculptures, created art that can move from one place to another using wind as its mobilizer.

A third milestone owes itself to the large-scale electronic information transmission network, the Internet, which gave rise to the last avant-garde movement of the twentieth century, *net.art*, whereby artists and activists subvert computer use and question both the machine and the observer operating the machine and the network. The German philosopher, mathematician, physicist and publicist Max Bens, deepens the relation between the formal sciences, like computer science, with art, when he states that every appearance of a new style in art is linked to the introduction of mathematical methods or theorems.[3]

In its new social and institutional position, art is not completely understood nor is there agreement on its identity. Moreover, nowadays the arts must deal with issues such as the relation between artistic practice and academia or the forms of art circulation and finance. For some artists, the terms "academy" or "academic" are uncomfortable as they are inevitably associated with certain ways of thinking and making contemporary art. In general, artists have low salaries and they experience job instability, as they alternate between receiving some public funding and taking sporadic jobs in advertising, the entertainment industry, or other associated fields. Very few artists establish themselves in what are scarce academic positions, but when they do, they enjoy economic privilege and a network of collaboration. Generally, the entry of artists into academia contradicts the disruptive and dissident aspects inherent to contemporary art. Borgdorff (2012) argues that the

[2] For a deeper explanation see Oswald D (2022) Cybernetics, operations research and information theory at the Ulm School of Design and its influence on Latin America. *AI & Soc* 37, pp. 1045–1057.

[3] Max Bens explores the relation between art and mathematics in his book of 1949, *Die Mathematik in der Kunst* (*Mathematics in Art*).

performative power of (academic) theory competes with the performative power of the free practice of art. On the other hand, many artists are uninterested in sharing or teaching their creative processes, since these are hallmarks of their work and ensure it is recognised and is given social value.

Despite all these difficulties, enrolment in art schools has expanded over time. First undergraduate, and now masters, doctoral and postdoctoral programmes are available, inserting artistic research within the broad academy. Specialised journals can be found, like the *Journal for Artistic Research*, published by the Society for Artistic Research (SAR), and the model has opened the discussion of whether art should be viewed as a form of academic knowledge production, eminently inquisitive and questioning. In most art schools, individual work is promoted over collective work and works of art are not considered objects of knowledge, in the way in which papers and seminars are in science. Within art schools there are hegemonic powers that represent traditions that do not necessarily endorse new or alternative forms of art production. Regarding research infrastructure, the economy of means used by artists is always scarce.

In terms of the circulation of art, there are many institutions and people involved in its promotion, including art schools, theoreticians, critics, museums, galleries, biennials, festivals, curators and marketing systems. The art gallery is a common exhibition space, but very few artists enjoy a contract with a gallery that can provide them with a stable income. Especially in developing countries, artists must be not only creative but also savvy when writing proposals and organising their work life if they want to survive professionally.

Models for financing the artistic practice range from systems of public and private financing through grants to investment systems and speculators purchasing art. The public funding system seeks to ensure the quality of artistic projects and the quantitative impact on the public. However, mechanisms to ensure an in-depth evaluation of the work carried out by artists with respect to the policies proposed or implemented, are still lacking.

In terms of the art market, Bourdieu (1996) discusses the role of intermediaries in establishing the value of an artwork, noting the existence of conspicuous power struggles between them, while Becker (1974) is interested in the collective actions taken by intermediaries to establish themselves as economic entities. According to Masterworks,[4] the value of contemporary art as a capital investment over the period 1995–2021 was 13.5%, a higher rate than both one of the most important stock indices of the USA, Standard and Poor's (9.5%) and gold (5.9%).

More recently, even before the Covid-19 pandemic, many art spaces switched to virtual sales. In comparison to thirty years ago, the production and distribution of art over the Internet seems low cost (Deresiewicz 2021). But the question regarding who ultimately pays for the effort remains. Many platforms have hidden copyright and distribution rules that can affect the benefits accruing to artists (Buchholz et al. 2020).

While corporate interest groups and conglomerates may defend an appropriable art market, the value of the copy versus the original is a controversial issue. Walter Benjamin first problematized it in *The Work of Art in the Age of Mechanical Reproduction*, (*Das Kunstwerk im Zneitalter seiner technischen Reproduzierbarkeiti*) published in 1936 (2003). Recently, Winner (2016) has argued that the social pressure to judge authentic works of art as aesthetically superior and art reproductions as inferior has little or nothing to do with genuine aesthetic or critical judgement. From a historicist perspective, a forgery is visually indistinguishable from the original and operates strictly on our belief about who made the work, when and how. Currently, attempts to assess the value of artistic objects in digital (rather than mechanical) formats presents an interesting case. Non-fungible tokens (NFTs) are virtual objects that are supposedly irreplaceable, unique and non-replicable which allows them to be assigned a

[4] See the following link: https://www.masterworks.io/ Revised January 27, 2021.

digital "certificate of ownership" that is stamped on the artwork file they are contained in. However, as critics point out, this understanding stems from a fraudulent business model; ultimately this proof of ownership is meaningless since digital copies of the work can always be infinitely reproduced.

In this way, the art world presents unique characteristics beyond its expressive, historical and theoretical components, involving a long list of technological, socio-economic and academic dimensions, that may be seen as erratic or capricious by externals, with similarities and differences with the scientific world.

2.4 Why Integrate Art and Science?

In the last two centuries or so, the sciences have helped to increase life expectancy, feed billions of people and generate an almost infinite number of objects of practical and daily use. However, solving almost any complex private, public or global contemporary problem seems to require the participation of several disciplines. Thinking of science as a progressive advance towards a single truth can lead to unreasonable expectations, for example that uncertainty is transitory, and that one can wait for certainties to arrive before acting on complex problems.

Some authors from the Stanford school, including John Dupré, Ian Hacking, Peter Galison, Patrick Suppes and Nancy Cartwright, have argued that the phenomena studied by science can be addressed by disciplines other than science, in a polyphony of approaches and interpretations which they call *scientific plurality*. From this perspective, the integration of science into art enriches the reach and possibilities of science, vindicating a variety of forms of knowledge while advancing a new line of thinking about the meaning of scientific progress and its consequences. This aspiration might also contribute to disentangle science from capricious or hidden interests and, at the same time, subvert what Belting (1987) called the death of the history of art by freeing it from the market, the stagnation imposed by institutional disciplinary structures and, ultimately, from banality or insignificance. It could be a fertile field where impending chaos could bear fruit to an infinity of possibilities (Prigogine 1996).

An interpretative openness and critical reflection about scientific objectivity can be found in examples of artistic representation such as Joaquín Torres García's inverted map of South America (1943) which reconfigures the socio-political and cultural context of the "south", revealing how the forms of representation, in this case geographical, may suggest geopolitical, hierarchical relations. Both art and science use perceptual cues (mainly visual and auditory ones) to emphasise and shape representations. At the same time, both make manifest empirical relationships, figurative models and abstract theoretical relationships as, for instance, in botanical and zoological atlases. In this sense of representation, art produces knowledge (Sullivan 2010; Savin-Baden and Wimpenny 2014) as a form of epistemology that transcends aesthetic experience (Eisner 2006), as an interpretation of the meaning of things (Barthes 2009) or through the anthropological interpretation of cultural objects (Gell 2016).

In opposition to a contemporary art that is immersed in postmodern, market-driven culture, some artists are conceiving and creating art as a domain of research, whereby questions or working hypotheses are key.[5] They claim that in this way art can provide us with valuable reflexive material, while simultaneously transforming how it circulates. As will be presented in Chap. 4, some artists have explored these experimental practices by integrating scientific domains into their art, thus generating a broad and emerging field of knowledge of still unimagined frontiers. A good example is the film *Primate Cinema, Apes as a Family* (2011). It is a work destined for both chimpanzees

[5] Learn on the relationship between research and artistic production through *Artist Based Research, Art Practice as Research* or *Art related Research* (Carter 2004; Sullivan 2005; Gray and Malins 2004; Hannula et al. 2005; Cazeaux 2008; Biggs and Karlsson 2010; Savin-Baden and Wimpenny 2014), which are based on a dual disciplinary background, as bioart or artificial intelligence, applied to artistic works.

and humans and presented on screens inside the chimpanzee pit at Edinburgh Zoo. Created by Rachel Mayeri and financed by the English Art Council, the work consists of a fictional story that captivates and entertains while triggering social interactions between captive chimpanzees and the humans visiting them. In the film, character actors appear who represent the dream of a chimpanzee living as a human being in a middle-class house. The film gives rise to an assortment of interesting questions: if primates dream, do they dream in a human way? Do humans and chimpanzees watching a fictional film together share in its meaning? What cinematic cues appeal to both monkeys and humans and what might this say about us and them?

The form of knowledge that emerges from art can be linked to the contextual view of scientific knowledge conceived by Rheinberger (1998), that is, as an experiential activity which combines local, technical, instrumental, institutional, social and epistemic elements. For Savin-Baden and Wimpenny (2014) the methods used by artists to collect data are different from those used in the hard sciences. For example, the types of instruments used by artists do not allow rigorous reproducibility. For Knowles and Cole (2008), artistic practice can be charged with emotional elements and affective experiences, wherein the imagination, as well as the intellect, provide ways of knowing and responding to the world. Lincoln and Denzin comment that arts-based research "involves processes of discovery and critical invention...(that) are moments of epiphany in culture suspended in time, for... the inspection and evaluation of institutions and their practices." (2003, p. 377).

In the variety of interpretations of artistic practice as research, Carter (2004) considers that rational discourse is materialised by bringing together disparate things (the cognitive, the phenomenological and the metaphysical). While Schwab (2013) focuses on the relation between experiments and artistic practice, comparing the laboratory with the studio, focusing on material practice, describing systems of creation and highlighting the temporal or experiential dimensions of production. Rheinberger's line of argument, discussed above, is useful for both scientists and artists when he writes that "experimentation is a site of familiarity, satisfaction and coercion" (Rheinberger 2021, p. 50), where experimental systems act as a "labyrinth under construction that dynamically redirects the viewer's gaze" (p. 51).

2.5 A Point of Encounter: Both Art and Science Rely on Metaphors

What makes it possible to distinguish an artistic object from an object that is not artistic? "Its poetics", the artist might say. And what is poetics? What is it that transforms an object into a piece of art?

In contemporary art, it is no longer possible to resort to an aesthetic or other sensible value independent of the concept (*conceptual art*) that justifies the work. In an artwork, there are several elements implied in its interpretation and judgement that are often not explicit, for example, the cultural moment in which the piece was conceptualised and produced.

When the piece of art is technically reproduced, what does the (indistinguishable) copy cease to have that the original retains? Walter Benjamin called it *aura*: its presence in time and space (2003). From an anthropological viewpoint, Alfred Gell (2016) also distinguishes an immaterial element present in artworks, namely, *agency*, as the intention assumed by the viewer behind the craft skills involved in the production of the piece.

The artistic piece does not have a practical function as long as such an art piece. Conversely, a functional object may become a piece of art, as in the *ready-mades,* created by Marcel Duchamp. In this artworks, the artist selects an existing object that is not artistic (a bicycle wheel, a spade, an urinary) and presents it in an expositive space (e.g. an art gallery) out of its practical context. When the artist "declares" that a piece is a piece of art, he/she adds the piece with an expressive intention that (may) produce new conceptual, interpretative or speculative relations in the observer. Viewed in this way, poetics is an articulation of metaphors, an expansion of significant meanings through a piece. This poetics may become explicit by a metalanguage that describes, analyses and classifies the work by, for example, an art critic.

Since Aristotle, poetics refers to metaphor to characterise the function and effectiveness of the art piece. For Ricoeur (2001) a metaphor is a trope (rhetorical figure) by resemblance which consists of a displacement that, through a substitution, achieves an extension in the meaning of the words. For Neumann-Soto (2012), a metaphor is a linguistic transposition or transfer. To say something about something using something different that emphasises one dimension of similitude but diverges in others: X is in A; A is in B; X is in B (Lakoff 1998). For example, in the phrase *The gold of the child's hair*, hair is related to gold in terms of colour, but not in terms of hardness.

Poetics, understood in its broadest sense, can then be found in any human activity as long as it admits a displaced or open reflexive approach. It appears when an object or activity allows us to distinguish a variety of open meanings or relations possible to ascribe to such an object or activity. Since Gell, all human activity can be referred to an agency (intention), and therefore to a metaphorical operation on its meaning. How to metaphorically refer to this (always speculative) intention is an open, free exercise. This metaphorical displacement could be more or less interesting or useful, and hence valued by a given culture. In this broad sense then, both arts and sciences are likely to be explored in terms of a poetics and link a metaphorical figure to its proposals.

The intentional divergence in the scope of an idea opened by a metaphor may refer to aspects of reality difficult to conceptualise (e.g. the concept of "energy"), or to create new mappings of the world (Padel 2011). Lakoff and Johnson (1980) give the metaphor a central role in language and cognition as they allow us to think the way we do. For them, concepts such as *state*, *cause*, *change*, *time* and *purpose* are exclusively metaphorical in nature. Metaphors are part of the conceptual system that our minds occupy to understand and process the world (Lakoff 2009). Their theory is closely related to the schemes of mental images and representations used in reasoning, as extension from bodily experiences of spatial and physical laws to more complex expressions. In an everyday example, "good" or "bad" mood could be expressed in terms of up-down directionality ("thumbs up"). It is the figure that is represented mentally. It is a characterization of knowledge that is constantly operating in the linguistic process (Schön 1966; Reddy 1969), which gives it an epistemological status.

Art basically performs metaphorical operations to generate or suggest relationships between ideas or concepts. The imagery of presence or absence of these relationships is used as a sign of the artwork (García 2014). A sign fixed in a representation may fall back on the world and society, and in so doing consolidates cultural icons. In science, we see that an explanatory contingency in a given moment in history has produced icons that have become cultural references of key achievements: the three-dimensional structure of atoms, complex molecules like DNA, fractal structures, the golden ratio, functional anatomy scans of the brain, electron and fluorescence micro photography and astronomical images are just a few amongst many examples.

Artists who stand on a meta position to science, may construct metaphors or imagery appropriate to the aspects of scientific research that may interest them. For example, they can "situate" scientific knowledge by metaphorically relating it to social, psychological or cultural dimensions, freed from the structure of scientific objectification that gave rise to them (Scott 2006). This operation of metaphorization ("artistification", poetic operation) of a scientific subject materialises in a piece (the artwork) without necessarily proposing new elements or associations that enrich scientific knowledge in that subject.

In science, metaphor is as ubiquitous as it is controversial (Haack 2019). Understanded as a diffuse rhetorical figure lacking a precise meaning, metaphors have been set aside by the sciences from their importance in the generation of knowledge (Kuhn 1993). However, solving scientific problems requires metaphoric representations to support reasoning and to propose explanatory models that underlie phenomena. The metaphor is, then, a transition to a new idea or theory. To understand how a natural entity "works", we need to metaphorize it as a machine (system) in order to represent the mechanism underlying the causal relationships expressed in the description. The interpretations of experimental results may converge into a meaningful metaphor (with its subjective imprinting) that culminates in a new model or theory. The use of metaphor by scientists could go beyond problem solving resulting in poetry as such, as have been extensively developed by Illingworth (2022).

Nowadays, scientific areas resort to conceptual metaphors in their terminology. For example, immunology employs metaphors associated with war: pathogens "attack", "invade" and "have strategies", while host bodies "defend" themselves. In biological evolution, living beings "adapt", increase or optimise their "fitness", and have "competitive advantage". In neuroscience, the brains, like computers, "capture" and "process information" to "make decisions". Even extra-terrestrial entities may turn into human-like "intelligent" beings able to contact us by using our current communication systems, as for example in Search for Extraterrestrial Intelligence, known as the SETI project.

Making visible and reflecting on these implicit metaphors opens the possibility of questioning or distancing ourselves from the premises these entail, enabling us to explore or enrich alternative perspectives. The problematization of metaphors is a powerful tool to inspire works of both art and science. When using a metaphor to describe scientific phenomena or theory, where is the limit of pertinence of it? How far may the war metaphors arrive, for example, in immunology? Or how open is a scientific field to accept new metaphors that may be as fruitful for understanding as those in use? Art, as a master practice in metaphor construction, may strongly impact the conceptual drift of science through art-science integration.

In this chapter we have briefly explored that both the sciences and the arts can be seen as social phenomena. Determined by a thickness of internal and external factors, science goes beyond truth and objectivity and arts goes beyond beauty and genius. Original, sometimes intriguing visions of both science and art have been presented, coming from perpendicular approaches feeded by anthropology, sociology, history and philosophy, that may hopefully open new perspectives on how these practices are understanded and approached. In the following chapter, the diversity of art-science manifestations is explored and some analytical cues are proposed to better understand this diversity.

Bibliography

Acord J (2009) Q&A: The art of transmutation. *Nature* 458, 577. https://doi.org/10.1038/458577a

Azimi R (2022) *Art market: Speculation grows around young artists.* Le Monde. https://www.lemonde.fr/en/money-investments/article/2022/12/28/art-market-speculation-around-young-artists_6009446_102.html

Bachelard G (1984) *The New Scientific Spirit.* Trans. Arthur Goldhammer. Boston: Beacon Press.

Barthes R (2009) *La aventura semiológica.* Madrid, Paidós.

Becker H (1974) Art as Collective Action. *American Sociological Review* 39 (6): 767-76. https://doi.org/10.2307/2094151

Becker H (1982) *Art Worlds.* Berkeley and Los Angeles: University of California Press, cited in Kimmel, M. S. (1983). [Review of Art Worlds., by H. S. Becker]. *American Journal of Sociology*, 89(3), 733–735. http://www.jstor.org/stable/2779026

Becker H (2011) *Manual de Escritura para Científicos Sociales.* Siglo XXI Editores. Buenos Aires.

Belting H (1987) *The End of the History of Art?.* University of Chicago Press.

Biggs M and Büchler D (2010) "Communities, Values, Conventions and Actions". In Biggs M and Karlsson H (eds.) *The Routledge Companion to Research in the Arts.* Londres: Routledge.

Borgdorff H (2012) "The Production of Knowledge in Artistic Research". In Biggs M and Karlsson H (eds.) *The Routledge Companion to Research in the Arts.* Londres: Routledge.

Bourdieu P (1996) *The Rules of Art: Genesis and Structure of the Literary Field.* Stanford, CA: Stanford University Press.

Boulton G, Bawa A, Babini D and Drury L (2021) Correspondence: Science community steps up to reform open access. *Nature* 599, pp. 556. https://doi.org/10.1038/d41586-021-03485-4

Buchholz L, Fine G A and Wohl H (2020) Art markets in crisis: how personal bonds and market subcultures mediate the effects of Covid-19. *Am J Cult Sociol* 8, pp. 462–476. https://doi.org/10.1057/s41290-020-00119-6

Carter P (2004). *Material Thinking: The Theory and Practice of Creative Research.* Melbourne: Melbourne University Publishing.

Cartwright N (1999) *The Dappled World.* Cambridge: Cambridge University Press.

Cazeaux C (2008) Inherently Interdisciplinary: Four perspectives on Practice-Based Research. *Journal of Visual Arts Practice* 7: 2, pp. 107–132.

Clark C (2021) *A new replication crisis: Research that is less likely to be true is cited more.* UC San Diego Today. https://today.ucsd.edu/story/a-new-replication-crisis-research-that-is-less-likely-be-true-is-cited-more. Reviewed February 23, 2023.

Deresiewicz W (2021) *La Muerte del Artista. Cómo los creadores luchan por sobrevivir en la era de los billonarios y la tecnología.* Capitán Swing.

Dupuy J P & Varela F (1995) "Circularidades creativas: para la comprensión de los orígenes". In Watzlawick Paul & Krieg Peter (comps.) *El ojo del observador.* Gedisa Editores.

Eisner E (2006) Does arts-based research have a future? Inaugural lecture for the first European. Conference on Arts-based research: Belfast, Northern Ireland, June 2005. *Studies in Art Education*, 43 (1), 9–18.

Eveleth R (2014) *Academics Write Papers Arguing Over How Many People Read (And Cite) Their Papers.* Smithsonian Magazine. March 25, 2014. https://www.smithsonianmag.com/smart-news/half-academic-studies-are-never-read-more-three-people-180950222/. Reviewed January 25, 2023.

Garcia T (2014) *Form and Object. A Treatise on Things.* Edinburgh University Press.

García-Bullé S (2021 April 27) *What is Replicability, and Why is it in Crisis?* Observatory Institute for the Future Education. https://observatory.tec.mx/edu-news/replication-crisis/

Gell A (2016) *Arte y Agencia. Una Teoría Antropológica.* SB Editores, Buenos Aires.

Gibbons M (1999) Science's new social contract with society. *Nature* 402 (Suppl 6761), C81–C84. https://doi.org/10.1038/35011576

Gosselin R D (2020) Statistical Analysis Must Improve to Address the Reproducibility Crisis: The Access to Transparent Statistics (ACTS) Call to Action. *BioEssays.* 42 (1): e1900189 https://doi.org/10.1002/bies.201900189

Haack S (2019) The Art of Scientific Metaphors. *Revista Portuguesa de Filosofia* 75 (4):2049–2066. https://doi.org/10.17990/rpf/2019_75_4_2049.

Haelle T (2021) *A massive 8-year effort finds that much cancer research can't be replicated.* Science News. https://www.sciencenews.org/article/cancer-biology-studies-research-replication-reproducibility. Review October, 2023

Hannula M, Suoranta J and Vadén T (2005) *Artistic Research: Theories, Methods and Practices.* Gothenburg: University of Gothenburg.

Haraway D (2019) *Seguir con el Problema.* Consoni.

Howe N (2020) 'Stick to the science': when science gets political. *Nature.* Podcast in https://www.nature.com/articles/d41586-020-03067-w. Reviewed January 25, 2023.

Illingworth S (2022) *Science Communication Through Poetry.* Springer.

Kellert S H, Longino H and Waters K C (2006) *Scientific Pluralism. Studies in Philosophy of Science Vol XIX.* Minnesota University.

Kuhn T S (1993) "Metaphor in science". In *Metaphor and Thought*, pp. 533 - 542. Edited by Ortony A Cambridge University Press.

Knowles G J and Cole A (2007) *Handbook of the Arts in Qualitative Research. Perspectives, Methodologies, Examples, and Issues.* Londres: Sage Publications.

Kozlov M (2023) 'Disruptive' science has declined - and no one knows why. *Nature* 613, 225. https://doi.org/10.1038/d41586-022-04577-5

Ladd J M, Lappé M D, McCormick J B, Boyce A M and Cho M K (2009) The "how" and "whys" of research: life scientists' views of accountability. *J Med Ethics.* December ; 35(12): 762–767. doi:https://doi.org/10.1136/jme.2009.031781.

Lakoff G and Johnson M (1980) *Metaphors we Live By.* Chicago, Chicago University Press.

Lakoff G (1998) "The contemporary theory of metaphor". In *Metaphor and Thought.* Ortony A. Cambridge University Press.

Lakoff G (2009) The Neural Theory of Metaphor. In Gibbs R (2008) *The Metaphor Handbook*, Cambridge University Press.

Latour B and Woolgar S (1995) *La vida en el laboratorio. La construcción de los hechos científicos.* Alianza Editorial. Madrid.

Latour B (1992) *Ciencia en acción. Cómo seguir a los científicos e ingenieros a través de la sociedad.* Barcelona: Labor.

Latour B (2004) *Politics of Nature: How to Bring Science into Democracy.* Translated by Porter Catherine. Cambridge, MA: Harvard University Press.

Latour B (2007) *Reassembling the Social: An Introduction to Actor-Network-Theory: An Introduction to Actor-Network-Theory.* OUP Oxford.

Latour B (2010) *Crónicas de un amante de las ciencias.* 1a ed. Buenos Aires: Dedalus, Colección Ciencias Humanas.

Ledford H (2019) US trust in scientists is now on par with the military. *Nature.* https://doi.org/10.1038/d41586-019-02389-8.

Lehrer J (2010) *The Truth Wears Off.* The New Yorker. https://www.newyorker.com/magazine/2010/12/13/the-truth-wears-off. 13 December, Reviewed January 25, 2023.

Lincoln Y S and Denzin N (2003) “The Revolution in presentation”. In Lincoln Y S and Denzin N K (eds). *Turning Points in Qualitative Research. Tying knots in a handkerchief.* Walnut Creek CA: AltaMira Press.

Longino H (1990) *Science as a Social Knowledge.* Princeton University Press

Ludwig D and Ruphy S (2021) Scientific Pluralism. *The Stanford Encyclopedia of Philosophy* (Winter Edition), Edward N. Zalta (ed.).

McCormick J B, Boyce A M, Ladd J M and Cho M K (2012) Barriers to Considering Ethical and Societal Implications of Research: Perceptions of Life Scientists. *AJOB Primary Research* 3(3), 40-50. https://doi.org/10.1080/21507716.2012.680651

Meillassoux Q (2020) *Metafísica y Ficción Extracientífica*. Editorial Roneo.

Munafò M R, Nosek B A, Bishop D V M, Button K S, Chambers C D, du Sert N P, Simonsohn U, Wagenmakers E J, Ware J J and Ioannidis J P A (2017) A manifesto for reproducible science. *Nat Hum Behav*. Jan 10;1:0021. https://doi.org/10.1038/s41562-016-0021.

Nature Editorial (2019a) The mental health of PhD researchers demands urgent attention. *Nature* 575, 257–258 (2019) https://doi.org/10.1038/d41586-019-03489-1

Nature Editorial (2019b) Scientists must rise above politics - and restate their value to society. *Nature* 572, https://doi.org/10.1038/d41586-019-02379-w

Neumann-Soto H (2012) La Metáfora en Aristóteles y su Contribución a la Comprensión. *Byzantion nea hellás*, (31), 47-68. https://doi.org/10.4067/S0718-84712012000100003.

Nieto Olarte M (n.d.) Historia de la Ciencia. Bruno Latour, Apuntes de Clase. https://historiadelaciencia-mnieto.uniandes.edu.co/pdf/BRUNOLATOUR.pdf. Reviewed January 24, 2022.

Padel R (2011) *The science of poetry, the poetry of science*. The Guardian. https://www.theguardian.com/books/2011/dec/09/ruth-padel-science-poetry

Park M, Leahey E and Funk R J (2023) Papers and patents are becoming less disruptive over time. *Nature* 613, 138–144. https://doi.org/10.1038/s41586-022-05543-x.

Prigogine I (1996) *El Fin de las Certidumbres*. Santiago de Chile: Editorial Andrés Bello.

Reddy M J (1969) “A Semantic Approach to Metaphor.” In *Papers form the ninth regional meetings*. Chicago Linguistic Society. Chicago University Press, Department of Linguistic.

Rheinberger H-J (1998) “Experimental Systems, Graphematic Spaces”. In *Inscribing Science: Scientific Texts and the Materiality of Communication,* edited by Timothy Lenoir, 285–303. Stanford, CA: Stanford University Press.

Rheinberger H-J (2021) *Iteraciones*. Editorial Pólvora. Santiago de Chile.

Ricoeur P (2001) *La metafora viva.* Translated by Agustín Neira. 2nd Edition. Ediciones Cristiandad & Editorial Trotta, Madrid.

Rojas S (2012) *El Arte Agotado.* Editorial Sangría.

Saltelli A and Funtowicz S (2017) What is science’s crisis really about? *Futures*, Volume 91, pp. 5-11. https://doi.org/10.1016/j.futures.2017.05.010

Sanderson K (2021) Researchers voice dismay at all-male science Nobels *Nature (News).* https://doi.org/10.1038/d41586-021-02782-2

Savin-Baden M and Wimpenny K (2014) *A Practical Guide to Arts-Related Research.* Rotterdam: Sense Publishers.

Schaeffer JM (2005) *Adiós a la estética.* Translation by Javier Hernández. A. Machado Libros S. A.

Schwab M (ed.) (2013) *Experimental Systems Future Knowledge in Artistic Research.* Lovaina: Universitaire Pers Leuven.

Scott J (2006) *Artists in Lab. Process of Inquiry.* Springer; Auflage.

Serra-Garcia M and Gneezy U (2021) Nonreplicable publications are cited more than replicable ones. *Science Advances.* May 7 (21): eabd1705. https://doi.org/10.1126/sciadv.abd1705

Sevier L (2023) “The Treachery of Publications”. Jan 8 https://www.researchtheory.org/post/the-treachery-of-publications. Cited by Flaherty C (2023) in *The Decline of ‘Disruptive’ Science.* January 17. https://www.insidehighered.com/news/2023/01/17/study.

Sormani P, Carbone G and Gisler P (2019) *Practising Art / Science. Experiments in an Emerging Field.* Routledge Advances in Sociology.

Schön D (1963) *Displacement of Concepts.* New York. Humanities Press.

Stroffolino A (2019) *The scientist’s dilemma: the ethics of advocacy.* A dissertation submitted to the School of Graduate Studies Rutgers, The State University of New Jersey.

Sullivan G (2009) *Art Practice as Research. Inquiry in Visual Arts.* Sage Publications.

Suppes P (1978) The Plurality of Science. PSA 1978: *Proceedings of the 1978 Biennial Meeting of the Philosophy of Science Association. Vol. II.* Eds. Asquith P and Hacking I. East Lansing: Philosophy of Science Association, pp. 3–16.

Van Noorden R (2017) The science that’s never been cited. *Nature* 552, pp. 162-164. https://doi.org/10.1038/d41586-017-08404-0

Gallup (2019) *First Wave Findings. How does the world feel about science and health?* Wellcome Global https://resources.informalscience.org/wellcome-global-monitor-how-does-world-feel-about-science-and-health. Reviewed February 23, 2023.

Winner E (2016) Ways of seeing. *Aeon.* https://aeon.co/essays/how-experimental-psychology-can-help-us-understand-art. Reviewed January 16, 2022.

Woolston C (2021) Stagnating salaries present hurdles to career satisfaction. *Nature* 599, pp. 519-521. https://doi.org/10.1038/d41586-021-03041-0

3 The Diversity of Art-Science Integrations

Abstract

Illustrated with full colour images, this chapter presents a detailed review of art-science manifestations that display diverse purposes and were produced in different historical periods and contexts. An analytical framework is then proposed to help us understand how art-science integrations move from the authority and predominance of one or the other discipline (art or science) using different modes of circulation depending on the purpose of the work. Finally, a categorization of art-science works that exhibit this spectrum of purposes is proposed and illustrated with examples.

Keywords

Art-Science · Integrated Curriculum · Interdisciplinary method · Teaching workshops

Having discussed in the previous chapter the challenges and limitations of artistic and scientific practice, this chapter explores different ways in which the sciences and the arts have become integrated as a broad and diverse space of dialogue and tension. The first section presents some historical antecedents of art-science integration and highlights some emblematic art-science works that have arisen in a diversity of contexts, both historical and geographical. Then, to further our understanding of art-science integration as such, an analytical framework and categorization is proposed, the aim of which is to illuminate the examples discussed as well as provide a heuristic tool for research and investigation in the field.

3.1 Beginnings and Expansions of the Relationship Between the Arts and Sciences

For Platonists and Aristotelians alike, the natural world is multiple and varied. One object cannot be reduced to another because each obeys its own unique nature. With the rise of modern science founded by Kepler, Galileo, Descartes, Pascal and then Newton, this powerful set of beliefs was replaced by a new way of thinking, sometimes referred to as positivism, whereby measurement, objectivity and causality took hold as a way to explain nature in terms of mechanisms ruled by universal laws. In this way, the philosophy of the Renaissance recovered the pre-Socratic idea of the unity and homogeneity of all entities, where each of these is no longer explained by its inherent nature but rather by simple principles that are common to all. Concepts like the universal animation of nature, based on this basic simple rule, not

© The Author(s), under exclusive license to Springer Nature Switzerland AG 2024
M. Velasco and I. Nieto, *The Art-Science Symbiosis*, https://doi.org/10.1007/978-3-031-47404-0_3

only served as a referent for religion, magic and astrology, but also for proposing intrinsic laws of nature (Villoro 1993).

Early forms of mechanistic representation appeared in the East during the Middle Ages, for example, in *The Book of Knowledge of Ingenious Mechanical Devices* by Ismail al-Jazari (1136–1206) or later in Mansūr ibn Ilyās' (Fig. 3.1) *The Anatomy of the Human Body.*[1]

During the same period in the West, technical resources for making artistic representations originated from instruments more typically used for measuring distances according to Euclidean plane geometry, in disciplines like astronomy, cartography and architecture. Together with these measuring instruments, the development of perspective aid in the proportionate representation of objects during the Renaissance. Amongst many fine examples, it's worth noting those of the painter Giotto (1267–1336) and the architect Filippo Brunelleschi (1377–1476).

The theoretical and conceptual interest in the relationship between art and science can be traced back to the anatomical work of Mondino de' Luizzi (1270–1326) or Leon Battista Alberti (1404–1472). Alberti, an architect and mathematician, identified rules underlying artistic practice: body proportions, spatial perspective, colour or the importance of architectural design based on function, which gave rise to, for example, Leonardo's cosmogony (1452–1519). Most of these rules specified techniques that limited variations in the process of observing objects and in their recording and representation. This approach, which grants pride of place to empirical knowledge, generated causal explanations of phenomena that could be refined and adjusted, for example, to illustrate or reproduce complex systems. An example is the representations of the heliocentric model and the phases of the moon by Galileo Galilei (1564–1642) using a spyglass. Instruments were developed to observe how physical structures move over time or to showcase the applied potential of mathematics and physics, for example, the creation of the propeller by Leonardo or the drawing grid used by Albrecht Dürer. For Daston and Galison (2010) these feats of applied engineering had two conceptual consequences: they revealed the truth of nature in the name of objectivity; and provided seemingly irrefutable answers to questions of intellectual interest, granting these the status of truth or knowledge.

Later, Johann Wolfgang von Goethe (1739–1842), wrote a treatise on colour, *Zur Farbenlehre* (Theory of Colours). Joseph Priestley (1733–1804) created visualisations that marked the time lapses of his experiments. Equally remarkable are the drawings by Marie-Anne Pierrette (1758–1836) (Fig. 3.2) which identify different experimental objects and how they are used in the laboratory.

Another example of applied artistic representation in the nineteenth century is found in the work of the French scientist Claude Gay (1800–1873) (Fig. 3.3). In 1830 he was commissioned to identify and categorise the natural resources of a rising country which culminated in *Atlas of the Physical and Political History of Chile,* published in 1842. Composed primarily of a series of mineral, zoological and botanical illustrations, Gay selected for publication 315 plates from amongst more than 3,000 drawings he had sketched during his thirteen year-long journey in the country, some of which included representations of the mores and culture of the day. The *Atlas* is a stellar example of how a graphic work can portray the character of a country, while being deeply influenced by the artistic and cultural values of the time, as seen through the eyes and the authority of a foreign scientific figure.

At first this empirical way of answering questions was hermetic and individual, or else shared amongst small groups of inquirers. But by the end of the nineteenth century this investigative practice had greatly expanded, giving rise - amongst other examples - to chemistry laboratories as collective spaces for teaching and learning (Morris 2015). State-of-the-art technology was introduced in the design of these laboratories and in the provision of services and equipment, speeding up the experimental processes. The laboratories of scientists James Clerk Maxwell (1831–1879) and Thomas

[1] For further details see Alghamdi et al. (2016)

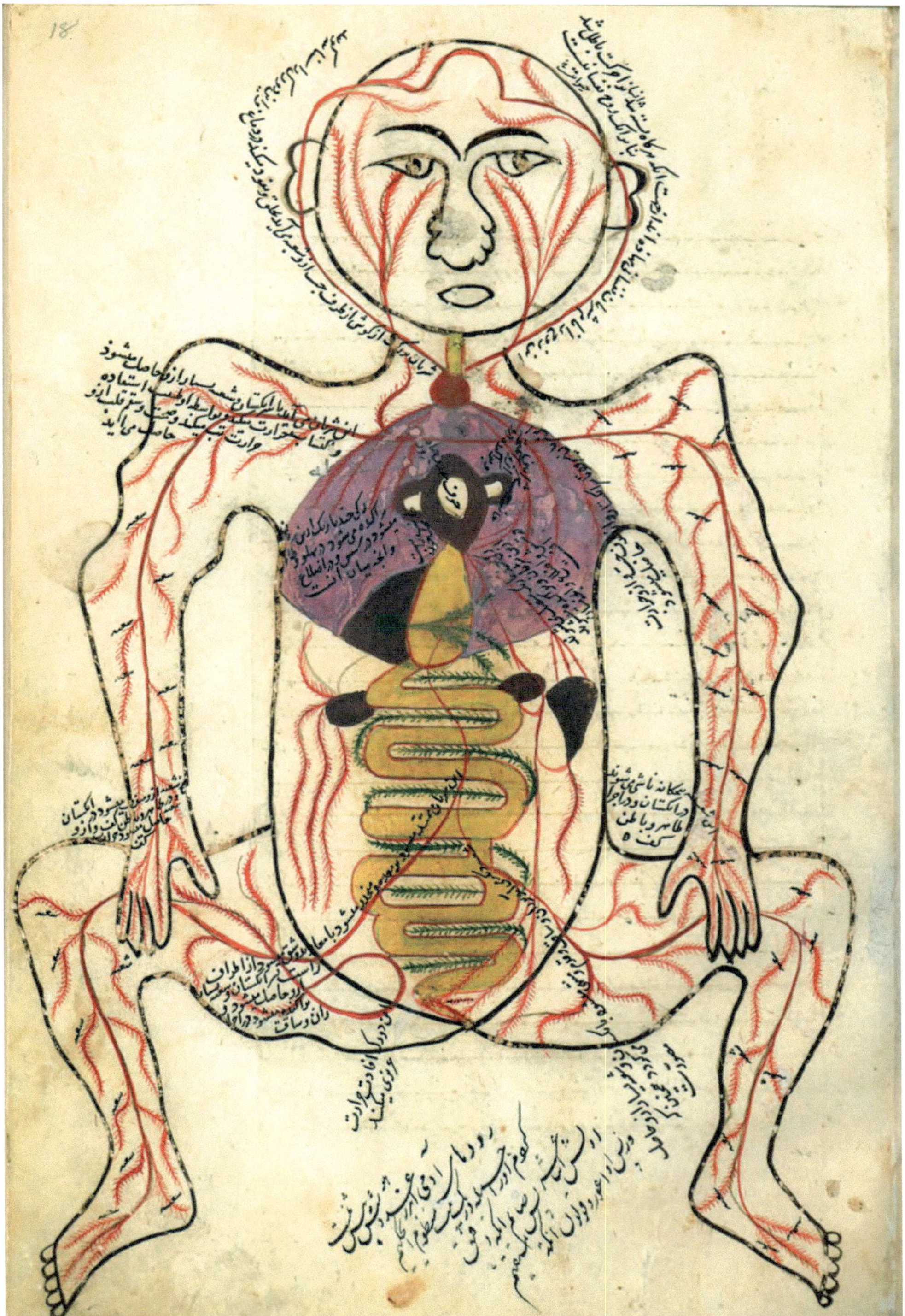

Fig. 3.1 Mansur ibn Ilyas (1390) Folio 18a from *Anatomy of the Human Body*. Courtesy of the The National Library of Medicine. https://collections.nlm.nih.gov/catalog/nlm:nlmuid-9406458-bk. Reviewed August 2023. Public domain

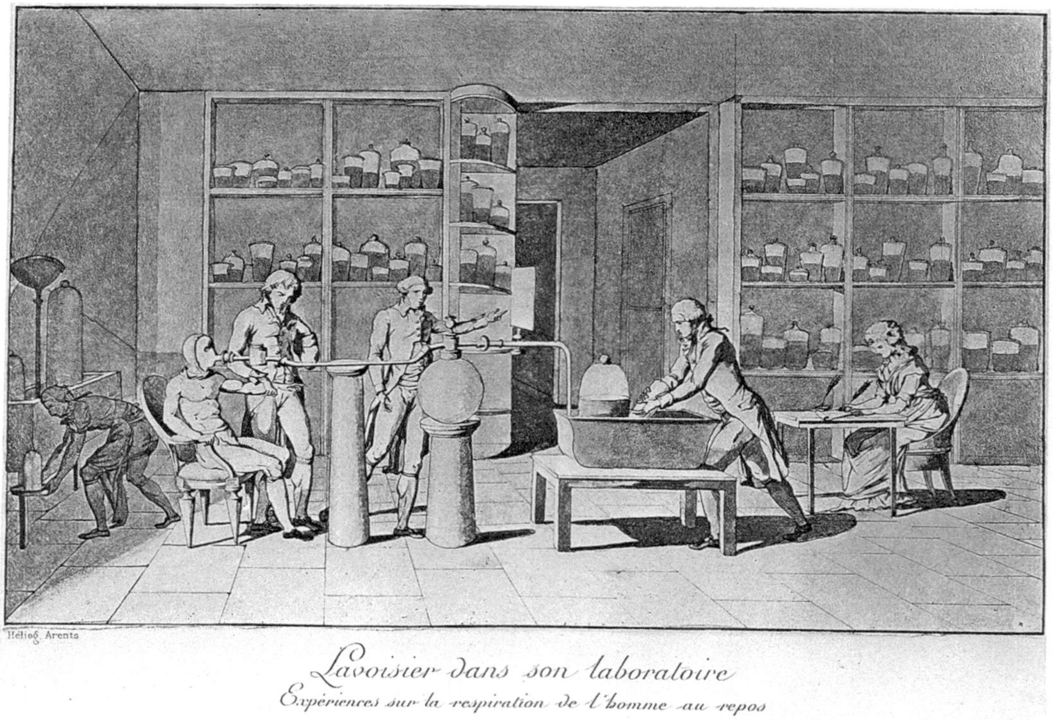

Fig. 3.2 Marie-Anne-Pierrette Lavoisier. Early 1790s. *Lavoisier experimenting on respiration.* Lavoisier experimenting on the respiration of a man at rest, with Lavoisier's wife, Marie-Anne Pierrette Paulze, taking notes on heliogravure. Lavoisier: Édouard Grimaux Published: 1888. Wellcome Library, London. http://wellcomeimages.org https://commons.wikimedia.org/wiki/File:Lavoisier_experimenting_on_respiration_Wellcome_L0000145.jpg. Reviewed January 29 2023. Public domain

Sutton (1831–1879) generated colour photography (Fig. 3.4) and the first sophisticated photographic devices used to study human and animal movement (Étienne Jules Marey 1830–1904) (Fig. 3.5).

In the twentieth century, the integration of art, design, science and technology spread to small groups after the wars and then settled in the academy. The Bauhaus School (1919) integrated art into industrial design; the Black Mountain College program (1933) integrated scientific subjects into artistic learning; and the Ülm School of Design (1953), where design was considered a systematic and scientific process, integrating into its curriculum interdisciplinary approaches such as cybernetics, widely developed by The Macy Conferences held in New York from 1941 to 1960.

Some other noteworthy precursors of hybrid art-science spaces are the *Experiments in Art and Technology* (E.A.T.) collective in 1967 in New York; the *Cybernetic Serendipity* exhibition in London in 1968 and the series of exhibitions called *Arte de Sistemas* (1969) organised by Jorge Glusberg, as new form of experimentation, expression and representation. Another important development was the creation of the journal *Leonardo* in Paris in 1968 by artist and engineer Frank Malina. It featured articles written by artists or theorists on crossovers between contemporary arts, sciences and new technologies. This journal continues today as a periodic peer-reviewed publication edited by Roger Malina and distributed by MIT Press. For further insight into the significance of scientific,

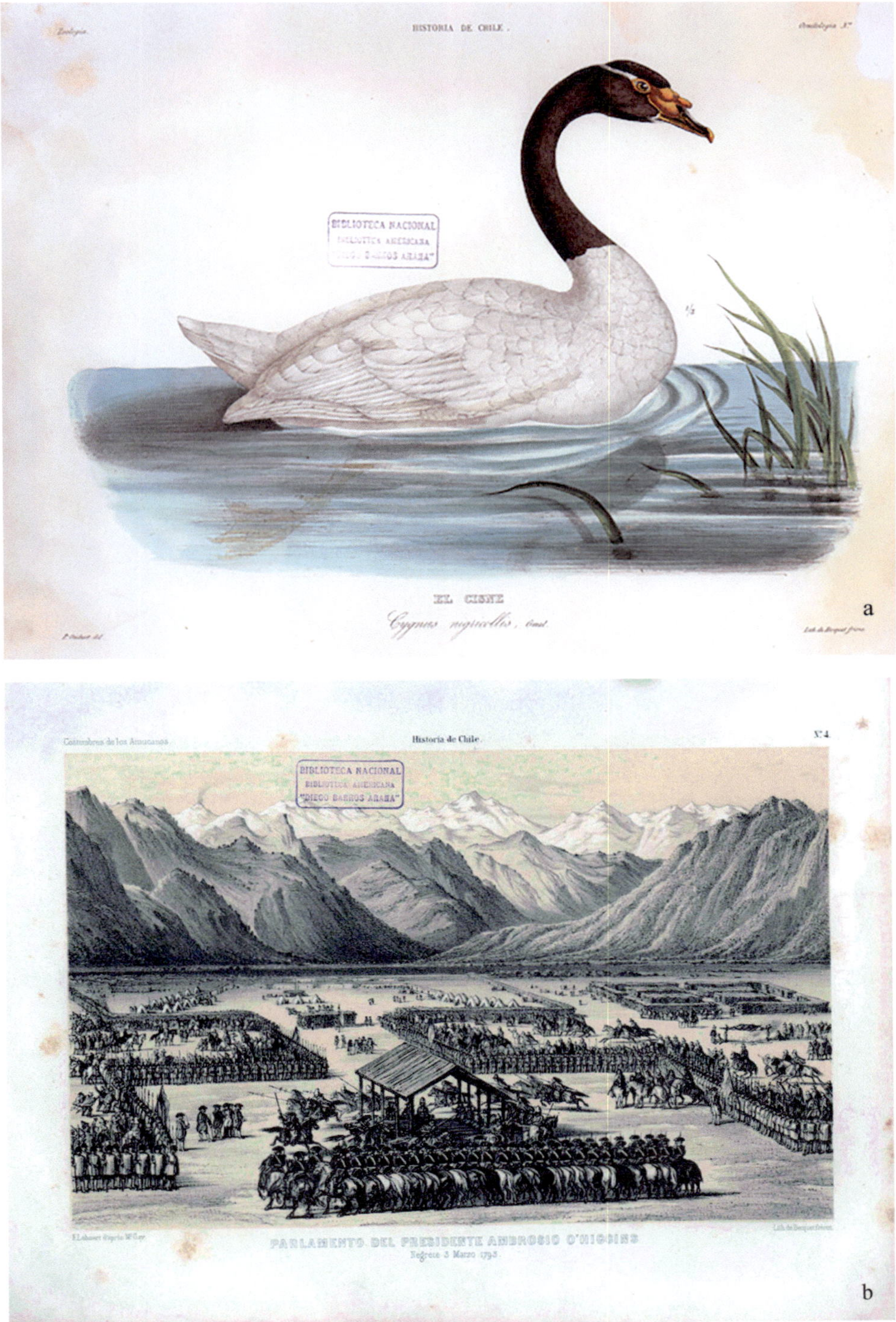

Fig. 3.3 Claude Gay. 1854. **a** *El Cisne. Cygnus nigricollis.* **b** *Parliament of the President Ambrosio O'Higgins.* In *Atlas de la historia física y política de Chile*. Volumes one and two, printed in E. Thunot, Paris. Memoria Chilena. Free to publish. https://www.memoriachilena.gob.cl/602/w3-article-8001.html

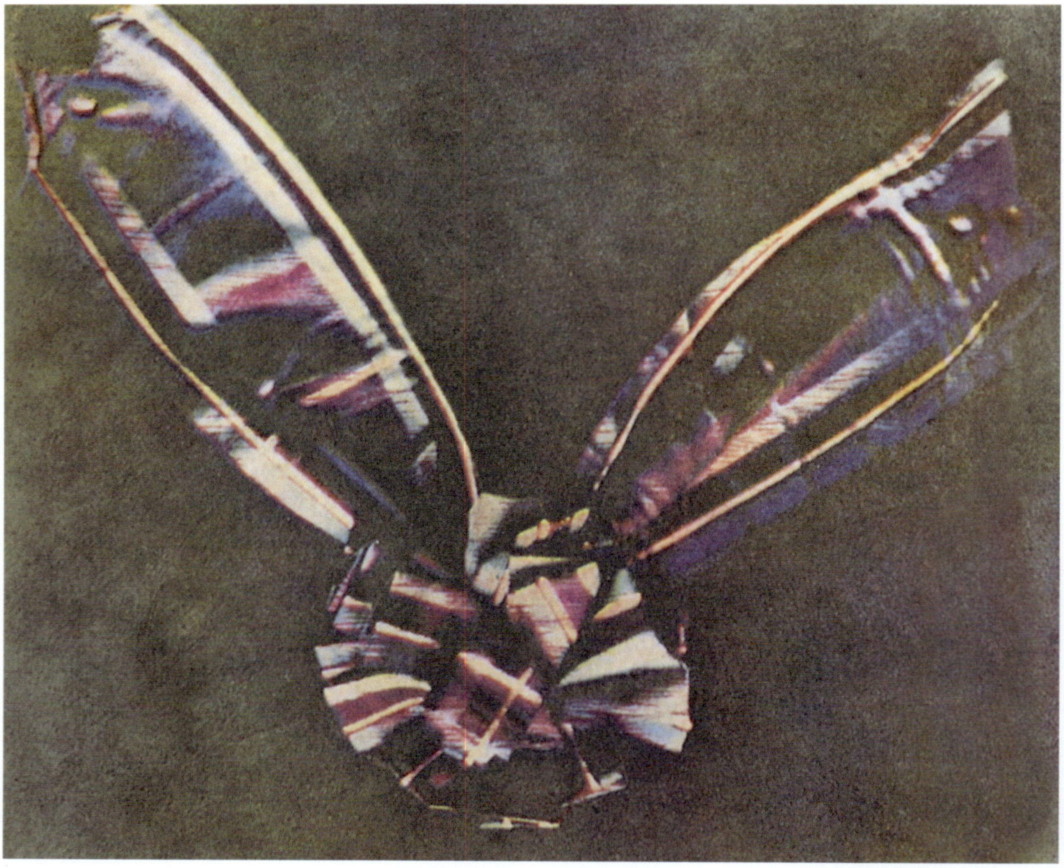

Fig. 3.4 James Clerk Maxwell. 1861. *Tartan Ribbon*. Photograph. Scanned from The Illustrated History of Colour Photography, Jack H. Coote, 1993. https://en.wikipedia.org/wiki/File:Tartan_Ribbon.jpg. Reviewed January 29 2023. Public domain

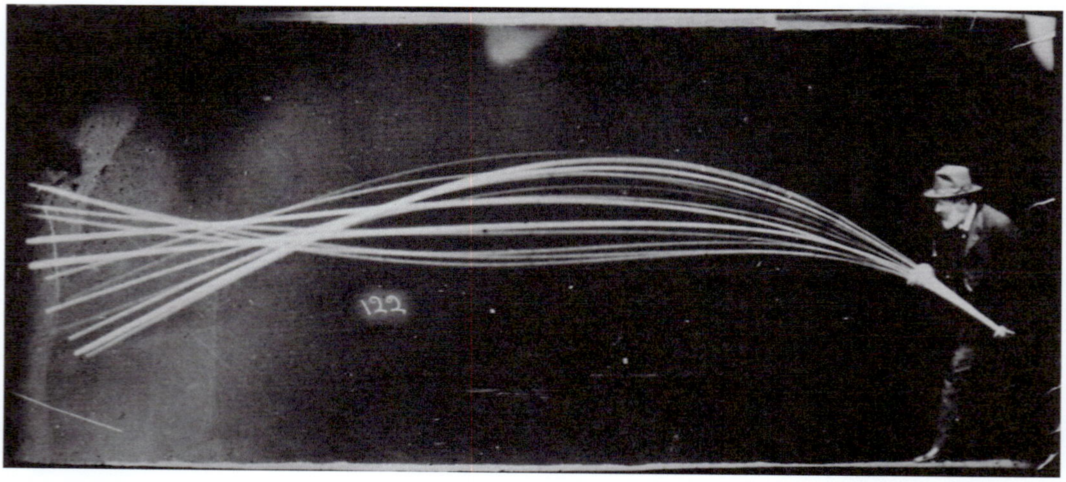

Fig. 3.5 Étienne-Jules Marey. 1886. *Marey shaking a flexible rod*. Photograph. https://commons.wikimedia.org/wiki/File:Marey,_der_einen_flexiblen_Stab_sch%C3%BCttelt_1886.jpg. Reviewed January 29 2023. Public domain

artistic and technological integration during the mid 1970, see.[2]

Nowadays the expansion of art-science practices into new fields is not just a topic of academic interest but also an important expressive resource for contemporary society. For example, *Media Labs* are research spaces where art, design, technology and media converge to undertake experimental work, teaching and the development of innovative technological products. Some of them focus their efforts on publicly driven initiatives, local demands for sustainable solutions, community collaboration and, crucially, inspiring collective critical thinking.

Our culture continues to be shaped by art-science integration which has become as ubiquitous as it is complex. The next section presents an analytical framework together with a proposal for a system of classification or categorization. Our aim in the development and elaboration of both is to better understand this art-science phenomenon and its impact today.

3.2 An Analytical Framework for Understanding Art-Science Integration

The integration of art and science can be understood in various ways, to begin, depending on the *purpose* behind the will to integrate both fields. The leader or author of an integrated work establishes a purpose and oversees its realisation, a process from which emerges a power relation with other participants, that is, the collaborating artist(s) or scientist(s).

Sometimes the purpose of the work is very clear in terms of the authority of one field over the other. In this case, an asymmetrical power relation between the two can be observed in the decision-making process. In other cases, the relation of authority between the two fields is the result of a negotiation process that gives rise to symmetrical power relations. It can be said that a symmetrical-asymmetrical polarity of authority frames a continuum of art-science manifestations in a pincer movement between the domain of the sciences and the arts. Its points of contact shift from one side to the other, denoting diffuse spaces, crossed limits and parallelisms. From here, a first criteria of analysis for art-science manifestations is identified:

Symmetric-asymmetric authority: *The relation of authority between the different disciplinary purposes involved and displayed in the art-science process.*

At the same time, the purpose behind the art-science work also defines the scope of communication platforms and representational choices in which the process and the results of integration are presented or exhibited. The scope of media platforms can be very broad or very specific, more accessible and open or more closed, depending on the intended audience and circulation strategy. Media platforms for art-science works include museums, public spaces, open talks, academic talks, press articles, peer reviewed articles, academic seminars and theses, movie and streaming platforms, video games, community workshops, disciplinary textbooks, art galleries, special interest societies, technical reports, experimental workshops, specialised software, patents, comic albums, posters, youtube channels, amongst many others. For example, a scientific model supported by the graphic skills of an artist may be published in a specialised journal but may also be presented in a TED talk. Or a photographer with a strong artistic purpose can capture an animal's unique behaviour, display it on his or her youtube channel while also collaborating on a scientific article with an animal behaviour scientist. This *media* aspect of an art-science integration results in a second criteria of analysis:

Open/Closed Media platform: *The way in which the process and the results of the art-science integration are communicated or exhibited through different formats and media.*

[2] UNESCO (1974). Impact of Science on Society. Vol. XXIV, No. 1, January-March. available at unesdoc.unesco.org/ark:/48223/pf0000009553.

Once the art-science integration work is communicated or exhibited, its form and content may impact and influence different fields: only the arts, only the sciences, both arts and sciences, or even other fields of knowledge. In this sense, the successful articulation of authority displayed by the purpose during the production process, and the choice of presentation media platforms adopted, will determine the realm and scope of the audience reached, the experience this audience will have, and hence, the potential influence of the work. If this influence is strong enough, a new disciplinary or interdisciplinary research process may be produced, starting a new cycle:

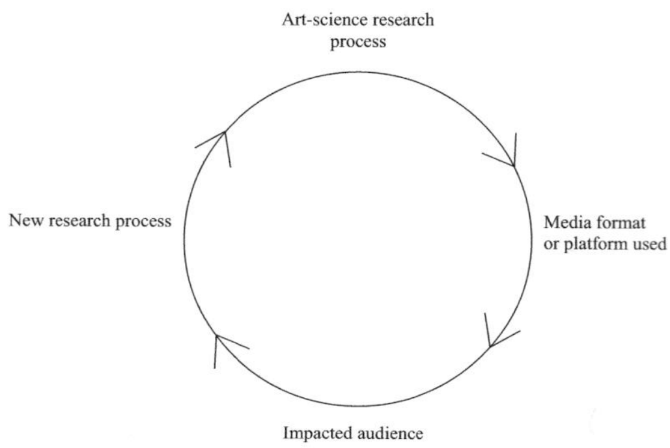

In a work whose purpose is purely scientific, the visual or spatial artistic influence follows strict guidelines derived from scientific objectives, basically concerned to support *scientific affirmations*. Scientific affirmations are understood here to mean a statement or sign that is true in accordance with evidence obtained in a replicable way. Artistic skills, thus, may influence or enrich the visualisation of mechanisms, models, input/output data or conclusive results. Visualisations function to provide a better understanding, description or generalisation of complex phenomena, and may be as diverse as the illustration of a botanical species, the representation of a virus, the trajectory of a mathematical function or the comparison between the size of rivers and mountains in different parts of the earth (Fig. 3.6).

In the chart presented in Fig. 3.6 made by Darton and Gardner in 1823, art is used as a means of presentation of the scientific measurement of different geographical bodies. In this case the primary purpose of the work is scientific, and it is science that holds the authority. The media platform of the chart is an engraving, apparently to be sold in Darton's own shop in London, rather than as part of a book or Atlas. This fact, together with its artistic and scientific value, defined its audience and scope of influence. In relation to this piece, Wiles, and then Tooley commented:

> The Enlightenment brought with it a new way of viewing the world. Mountains were no longer part of the earth but became discrete concepts that could be removed from their environment to be measured, defined and understood. During the 19th century, this resulted in comparison charts… showing the beauty that science could produce. (Wiles 2017, parr 1).
> This may be the earliest chart to depict the heights of mountains and the lengths of rivers together in one illustration, with the rivers hanging down over the mountains... This chart is beautifully done and, while not large in size, it is impressive in its scope. (Tooley 1979, p. 236)

In contrast, the scientific participation in a work whose purpose is purely artistic is subject to the *expressive/declarative* goals of the artist. Through these goals is that the artist inspires (or manipulates?) human emotions (Noah 2017). For Searle (1979), *expressive acts* occur when "we express our feelings and attitudes" (p. viii) such as when we "apologise, thank, condole, congratulate, complain, lament, protest, deplore, boast, compliment, praise, welcome, greet" (Searle and Vanderveken 1985, p. 211). On the other hand, *declarative acts* bring about changes in the world when successfully invoking an institutional or factual power (Searle and Vanderveken 1985). In contemporary art, this occurs when the artwork is successfully recognized as such by the *Art world* (Becker 1982). Here, the format or mode of scientific influence in the artwork is freely decided upon by the artist and may include theoretical content, images,

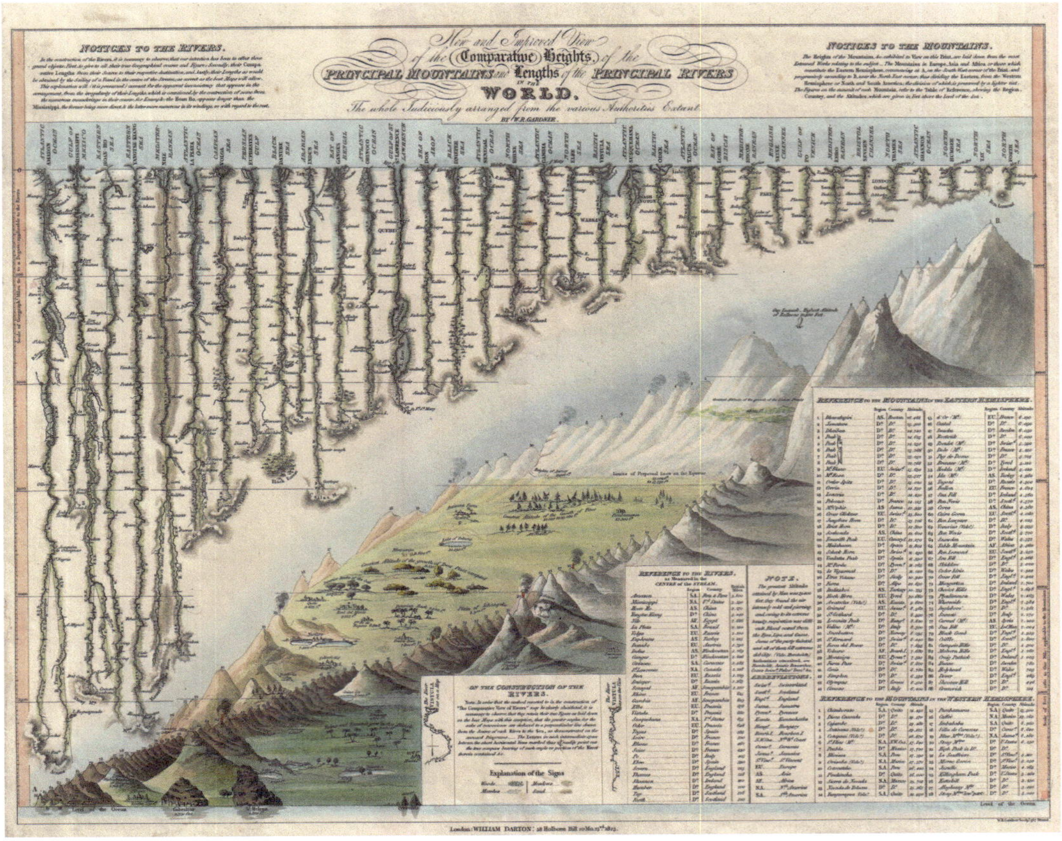

Fig. 3.6 William Darton and Wells Gardner. 1823. *Comparative Chart of World Mountains and Rivers—Geographicus—Mountainsand.* https://commons.wikimedia.org/wiki/File:1823_Darton_and_Gardner_Comparative_Chart_of_World_Mountains_and_Rivers_-_Geographicus_-_MountainsandRivers-darton-1823.jpg. Reviewed, January 29 2023. Public Domain

methods, instruments, personalities, materialities, even attitudes. The work may be speculative, uncertain or controversial as illustrated in Fig. 3.7.

> The *Semi-living Worry Dolls* is a project that reinterprets the story of the Guatemalan Worry Dolls to deal with contemporary anxieties. According to legend, Indigenous Guatemalans taught their children that they could solve their worries by whispering to their dolls each night before going to bed. Using this story as the starting point, *The Semi-living Worry Dolls* is an installation that allows audience members to express their fears about the impact of biotechnology in society and culture. Worry dolls were placed in the gallery and visitors could whisper their fears to each of the dolls. The dolls were hand-crafted using multiple cells—skin, muscle and bone cells—which were grown on a degradable polymer matrix and then stitched with surgical sutures. (Catts et al. 2001)

The purpose of this piece is clearly artistically driven, using sophisticated biotechnological resources necessary to expose, from the biotechnological material itself, the public's fear with respect to biotechnology research and products.[3], The texture, strength, colour and general appearance of living tissue supports worry dolls as part of an installation that is exhibited as an artistic experience. The discussion about biotechnology, nonetheless, has been addressed by the authors in different media platforms related to art, biotechnology, interdisciplinarity, biopolitics and bioethics reaching a wide audience and with significant impact.

[3] See Catts and Zurr (2014).

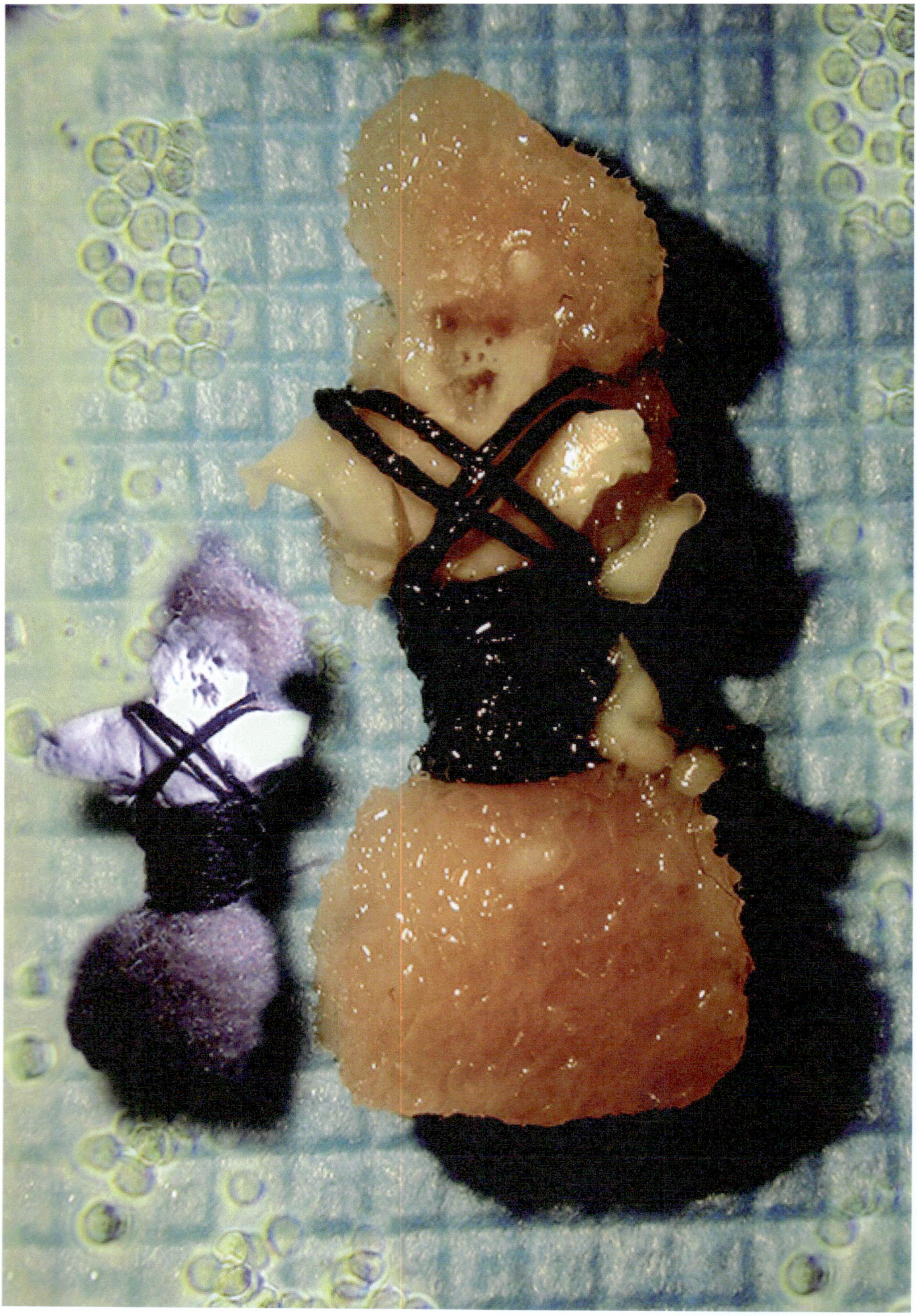

Fig. 3.7 Ionat Zurr and Oron Catts. 2000. *A Semi-Living Worry Doll H*. The Tissue Culture & Art Project. McCoy Cell line, Biodegradable/bioabsorbable Polymers and Surgical Sutures. The Tissue Culture & Art(ificial) Wombs Installation, Ars Electronica 2000. Image courtesy of the artists

The analytical framework presented above was conceived as a way to explore in a more formal way the art-science landscape. An important conclusion from the distinction of a kind of authority "gradient" among different art-science manifestations, is its consequences regarding how to establish proper and comparable validation criteria on the value and therefore the success of the work when an applied purpose is unclear. Both art and science have a framework or tradition on these validation criteria. However, interdisciplinarity in general, and art-science integration in particular, becomes a complex domain in these terms, in particular when the authority among the integrated disciplines is, at least, negotiated and therefore, in tension, as will be seen in Chaps. 4 and 5.

In the following, a categorization of some art-science works is presented. Some of these have a scientific purpose, others an artistic purpose, and still others fall in between, in an indeterminate space that allows for symmetrical integration. The reader is invited to examine and reflect upon these examples in terms of the relations of authority and the media platforms that are used.

3.3 Different Categories of Art-Science Manifestations

Identifying categories and classifying examples is always complicated. Here, our interest in doing so is to provide evidence of its richness and diversity of meaning and purposes of art-science manifestations. After reviewing a vast number of referential works, six categories have been identified.

(a) Artistic works aimed to honour, appreciate, disseminate and raise public awareness of the value and importance of science
(b) Works in which artistic skills are used to configure both visually and spatially, scientific theories, models or results
(c) Works that use science as a model or reference for art
(d) Works in which science is explored, problematized or questioned by art
(e) Works that provide a scientific explanation of the aesthetic experience
(f) Works that encourage and foster creativity and innovation by exploring the relation between the arts, technology and science.

As tends to occur with classificatory systems, the more cases are attempted to classify, the more come across outliers or anomalies. Put differently, some works of science-art integration will not neatly fit into any category while others might be admitted into more than one. That said, with its imperfections, the identification of categories is often useful as an instrument for ordering a field and deepening our understanding of it. The categories below have this intention and are described and illustrated with several examples. It may be worth noting that their order of presentation does not reflect a hierarchy of value or importance between the categories.

a. Artistic works aimed to honour, appreciate, disseminate and raise public awareness of the value and importance of science

Western art has represented its systems of beliefs through sculpture, painting and graphics, as well as installations, interventions of the body or the use of electronic media. These representations have variously idealised Western dogma, served the purpose of building or maintaining empire (and colonial subjugation) or have been used as part of a welfare-oriented education. Scientific topics are no strangers in these artistic representations. In more or less figurative and abstract modes, art has contributed to highlighting first the transcendent and then the positivist character of the specifically scientific way of knowing. The production of this type of artistic manifestation is materialised in various formats and platforms of presentation and representation, for example as monuments, paintings, prints, memorials, covers, educational models, catalogues or exhibition spaces.

With the transition from the Renaissance to the Enlightenment, a variety of representations of scientific activity begins to appear. Indeed, art plays a pivotal role in consolidating the empire of science over religious revelation for the pursuit of truth, gradually supplanting much of Christian iconography with new symbolic and institutional powers. The works in this category highlight a scientific celebrity or practice to honour their contribution to society. An impressive example is the almost allegorical cover of the pioneering book on human anatomy *Fabrica,* written by Andreas Vesalius (1514–1564) in 1543 (Fig. 3.8).

In the complex transitional stage between the acceptance of revealed truth and the emergence of empiricism, Athanasius Kirchner (1602–1680) applied the authority of the regularities of observed phenomena in nature to exalt a divine order of the universe. Dedicated to Emperor Ferdinand III, the work *Magnes sive de arte magnetica (1641)* elaborates on how the universe is ruled by universal forces of attraction and repulsion (Fig. 3.9). With its ultimate origin in God, all creation is attracted by the force of love toward him (Udías 2020). In this case, the authority belongs to faith and the natural regularities observed are functional to it. The medium is a book destined for a selected audience of that time and published with the resources made available from them.

From the seventeenth century onwards, a significant number of paintings depicting scientists and scientific activity were produced. One of the best known is the *Anatomy Lesson* by Rembrandt in 1632. It shows Dr. Nicolaes Tulp at work dissecting humans, together with a group of surgeons (who financed the painting) (Fig. 3.10). These dissections were rare and spectacular acts, to the point of becoming social events. They took place in conference rooms that were theatres. Such is the case of Waag in Holland, Padua in 1594 or Bologna in 1649.

Also, there are depictions of instrumental investigative work, such as the *Experiment with a Bird in an Air Pump* (1768) by Joseph Wright of Derby (Fig. 3.11).

During the twentieth century, representations of notable scientists can be found such as in the serigraph of *Albert Einstein* by Andy Warhol (1980).

Natural History Museums, institutions that are dedicated to the preservation, cataloguing and exhibition of specimens and often do research on the phylogeny and taxonomies of species, is another interesting example of art-science integration in this category. Their role as exhibition spaces, with all that this implies in terms of representation, is unquestionable. During the twentieth century, the exhibition and dissemination of knowledge unfolds through the representation of faithful explanatory or descriptive scientific models often created by artists and scientists and exhibited in spectacular centres, designed to facilitate and motivate the understanding, insertion and validation of science in societies that support it through public taxes (Stettler 2006).

Other examples of spaces that showcase science include the Exploratorium, founded in 1969 by Frank Oppenheimer in San Francisco, USA. Or public spaces used to pay homage to science, for example *The Logical Basis* (2016), an artwork that tributes to meteorologist and climatologist Syukuro Manabe made by British artist Liam Gillick and exhibited in the Gare du Nord station of the Paris Metro.

Contemporary science communication practice belongs to this category as a full-blown profession in its own right, informing, educating and raising awareness and wonder about science and science-related topics. Amongst other media, science communication employs documentaries, TV shows, podcasts, thematic books and periodical journals, all of which rely on artistic support, either on a permanent or an intermittent basis.

In a more corporate-driven scientific tribute, Eigler and Schweizer (1990) demonstrated the possibility of constructing images at the atomic

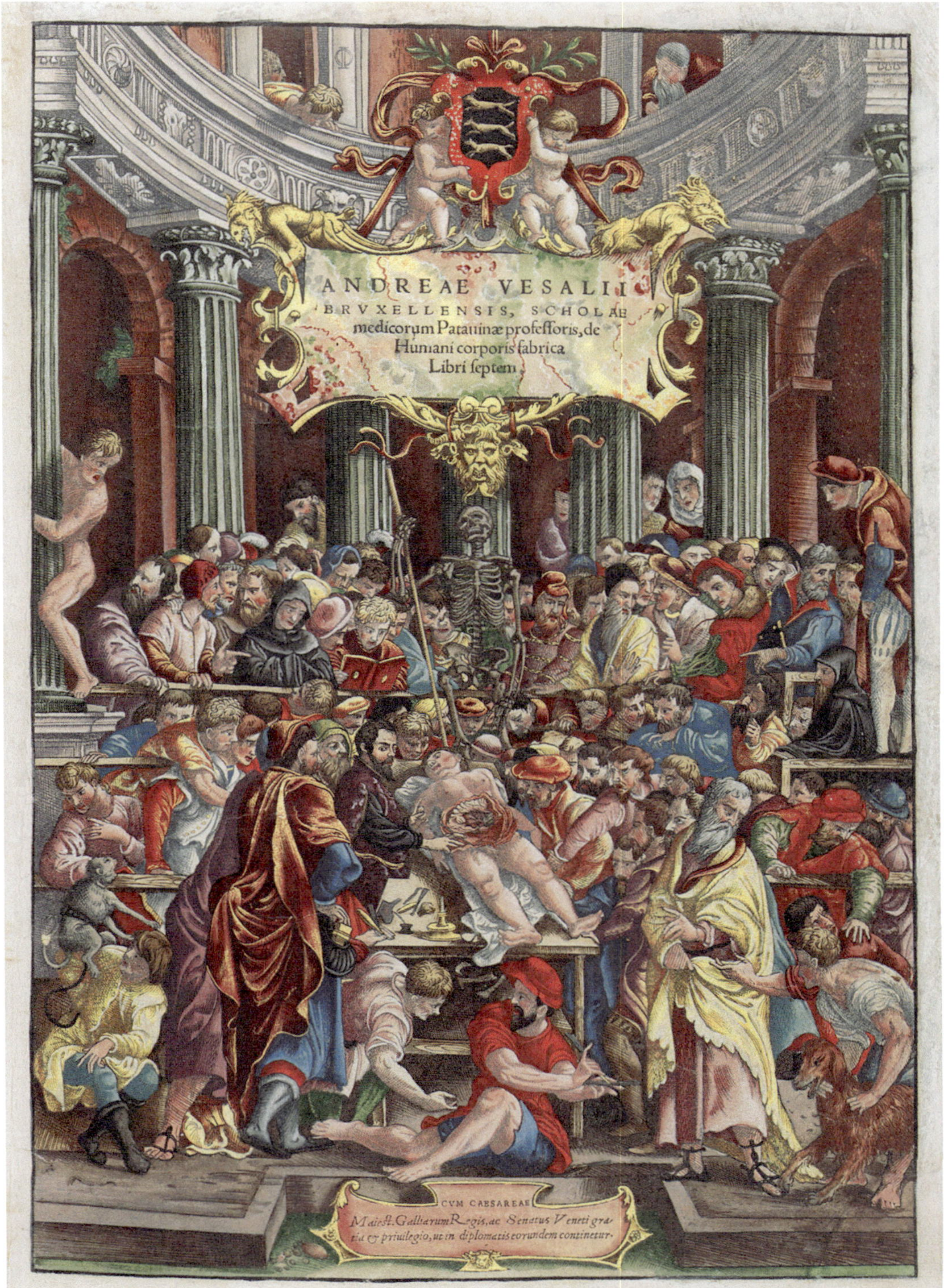

Fig. 3.8 Andreas Vesalius. 1543. *De humani corporis fabrica libri septem* (Book cover). Courtesy of an Anonymous Private Collector; photo courtesy of Christie's, New York Dedication copy presented by Vesalius to Charles V (1500–1558), Holy Roman Emperor. https://commons.wikimedia.org/wiki/File:Vesalius_color-1543.jpg. Reviewed January 29 2023. Public Domain

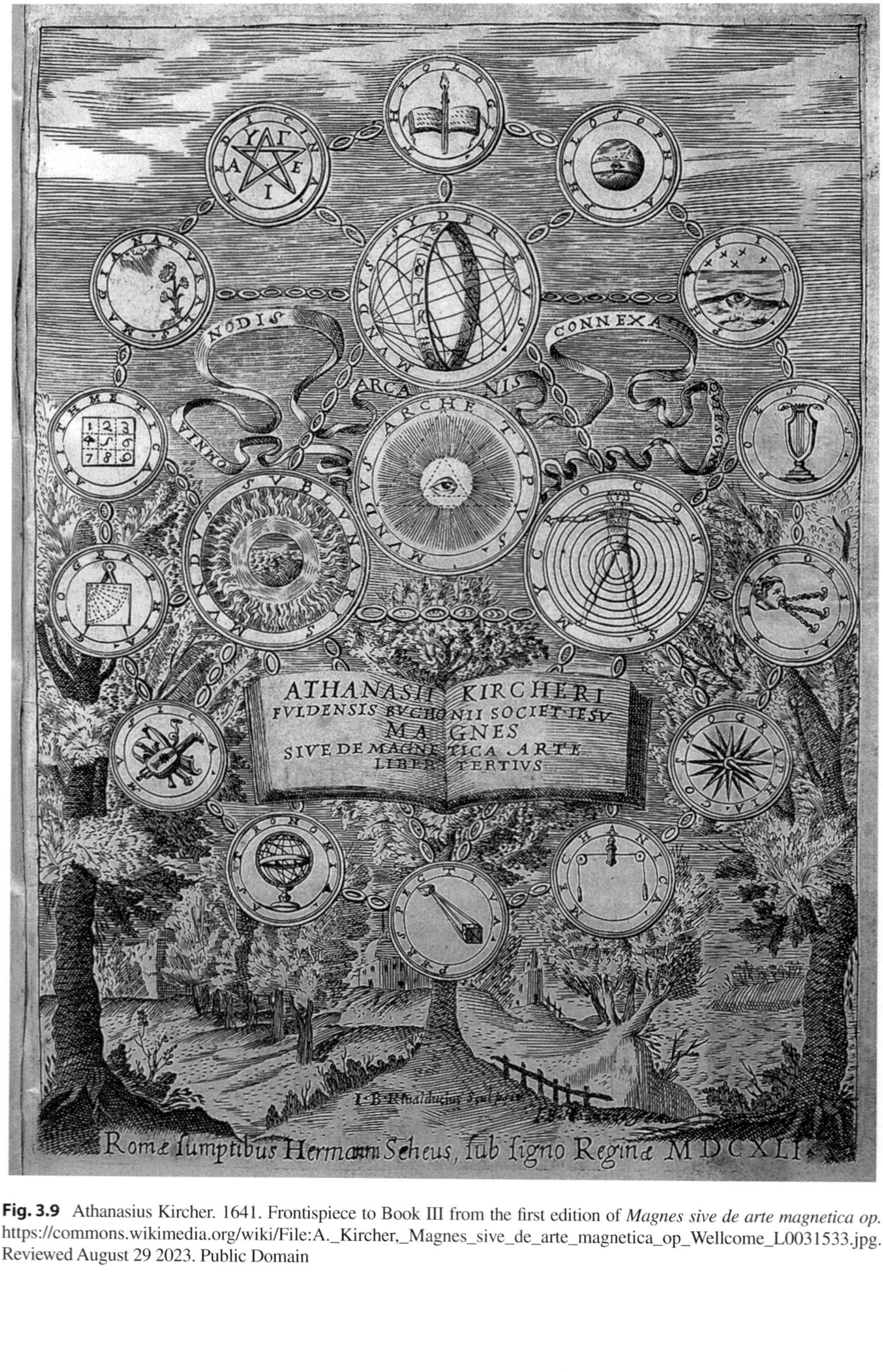

Fig. 3.9 Athanasius Kircher. 1641. Frontispiece to Book III from the first edition of *Magnes sive de arte magnetica op.* https://commons.wikimedia.org/wiki/File:A._Kircher,_Magnes_sive_de_arte_magnetica_op_Wellcome_L0031533.jpg. Reviewed August 29 2023. Public Domain

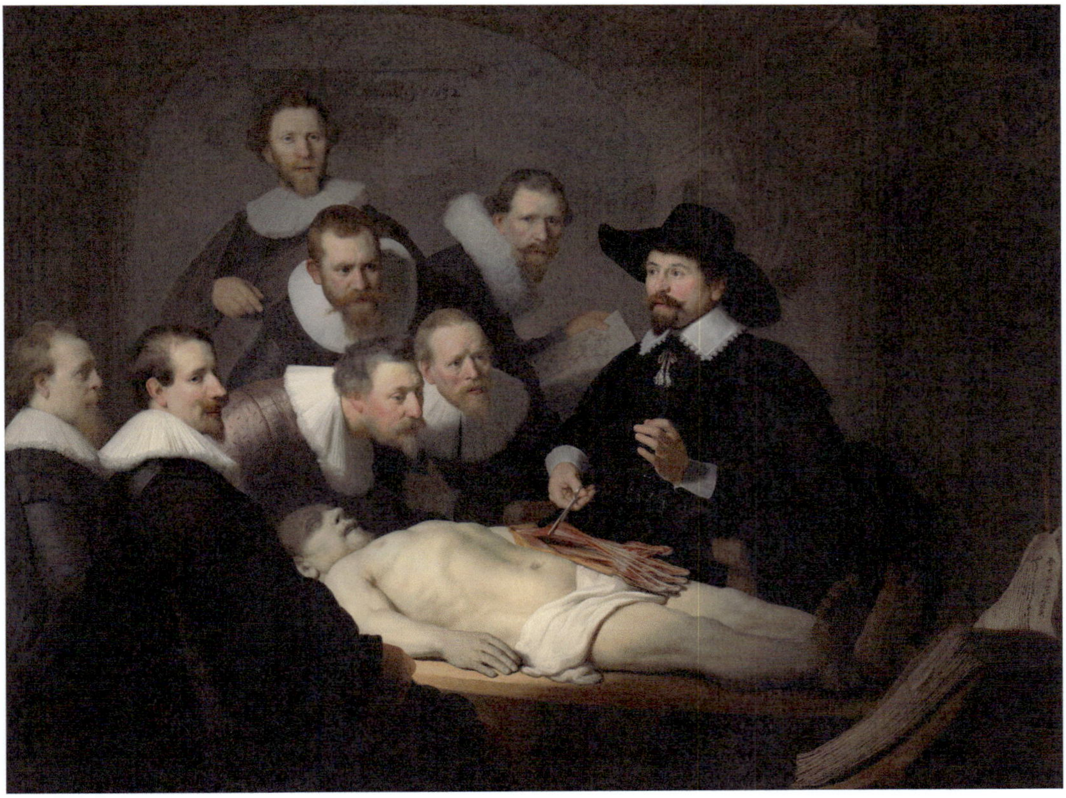

Fig. 3.10 Rembrandt Harmenszoon van Rijn. 1632. *The Anatomy Lesson of Dr. Nicolaes Tulp.* Photograph by Margareta Svensson, from the original painting https://en.wikipedia.org/wiki/File:Rembrandt_-_The_Anatomy_Lesson_of_Dr_Nicolaes_Tulp.jpg. Reviewed August 2023. Public Domain

level to spell out the name of one of their sponsors, IBM. Using single atoms, they demonstrated their capacity to control nanotechnology. With this kind of "signature" the researchers claim a certain closeness to artwork (Wildevuur 2009).

In this category, a power structure, private or public, usually provides the infrastructure and human and financial resources to highlight a scientific subject. We can say, then, that in this category, the criterion of scientific authority prevails as the origin of the valued content. However, it is the political authority which validates the tribute and exposes it publicly, just as a propaganda or ideological and religious manifestation would do. In this way, the artist's skills are in a position of asymmetry with respect to the political and/or scientific authority, who ultimately articulates or guides the final result, that is, the work of art. Interestingly, the image of the single atom IBM, was driven and published as a purely scientific work, but as Eigler claimed: "It was more than a publicity stunt. Emotionally, for me, it was much more important. This is going to sound hokey, but it's the truth. IBM picked me up off the scrap heap of science and gave me every opportunity a scientist could hope for to be successful… As far as I was concerned, it was payback time." (Shankland 2019, parr.8).

b. Works in which artistic skills are used to configure both visually and spatially, scientific theories, models or results

The interest in the aesthetics or visual impact of scientific images is neither accidental nor anecdotal. Scientists know that our fascination with images can enhance their results and motivate their teams (Kemp 2005). Currently,

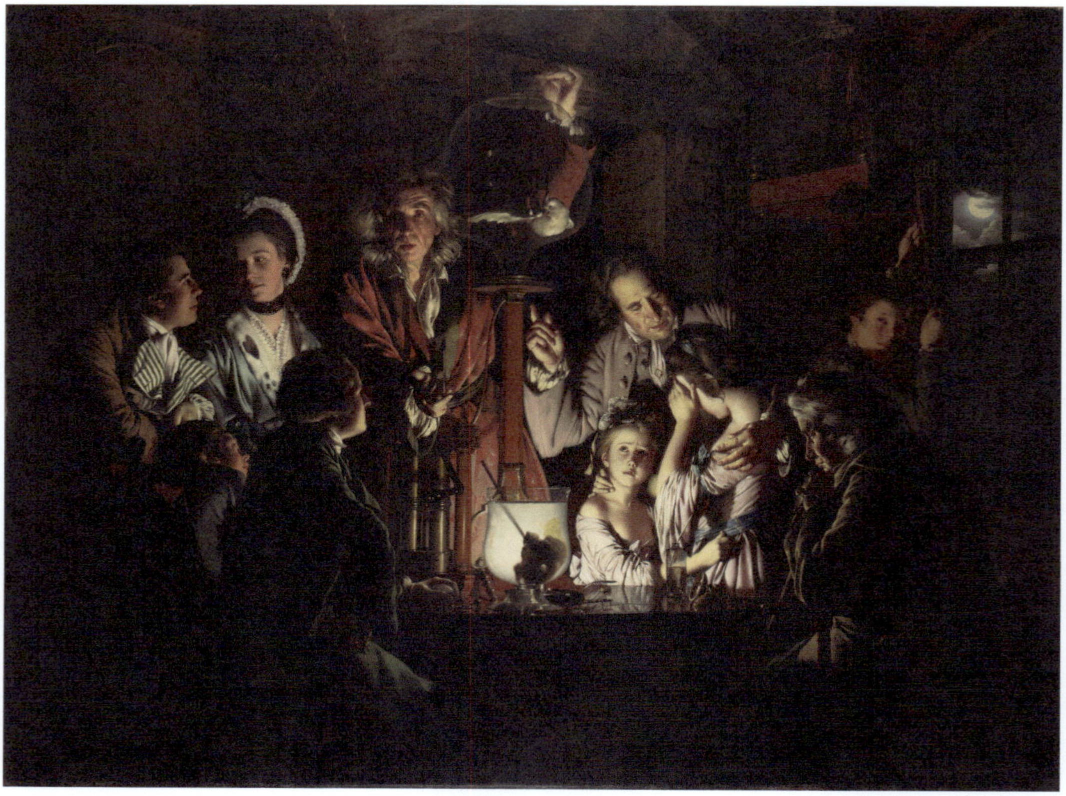

Fig. 3.11 Joseph Wright of Derby. 1768. *Experiment with a Bird in an Air Pump*. The National Gallery, London. https://commons.wikimedia.org/wiki/File:An_Experiment_on_a_Bird_in_an_Air_Pump_by_Joseph_Wright_of_Derby,_REVERSED.jpg. Reviewed January 30 2023. Public Domain

much scientific work consists of generating visualisations that constitute evidence or, in Latour's (1987) more conceptual terms, "inscriptions" of experimental or field evidence. Spectacular imaging is evident in areas such as cosmology and astronomy, and in electron and confocal microscopy.

When proposing visual evidence, why do scientists choose one image or model over another? Does this selective factor bias our understanding of a phenomenon? Since the origin of experimental and scientific activity, images have been used to present research results, and to illustrate concepts, models, hypotheses and theories. Scientists have employed art to describe pieces or specimens (mineral, archaeological, biological), anatomical, geographical or geological morphologies; or to create clearer, more attractive visualisations of cases, results, models and theoretical proposals. It may be thought that these visual devices are no more than an instrument in the service of science. In many cases, however, science draws upon commonplace or appealing visualisations (for their order, simplicity, symmetry, harmony) as heuristics that influence, inspire or even resolve a scientific hypothesis. According to Borel (1983), mathematics is an art in which aesthetic criteria also play an important role in the judgement of mathematical concepts. This suggests an epistemic role for aesthetic values in scientific practice, particularly in the formulation of questions, and in discovery and interpretation. Aesthetics may also influence the scientist in the case of underdetermination, for example when there is still no empirical confirmation of a theory (Ivanova 2017). In the past, this influence was decisive as the aesthetic virtue of models and theories was an indicator of its divine origin, and therefore, veritable (Norton 2020).

The artistic and naturalistic revolution of the Renaissance allowed the graphic illustration of both subjects and scientific procedures, giving rise to scientific practice accompanied by visual support. Originally, most scientific illustrators were traditional artists recruited for this work (Daston and Galison 2010). Examples of this practice can be found in the book *Anatomia humani corporis* by Govaert Bidloo (1685) which includes detailed anatomic illustrations made by artist Gerard de Lairesse; and the work of anatomist Paolo Mascagni (1755–1815). The scientific depiction of the anatomy of the human body involved complex representational challenges, at a time when artists were well trained in the detail of muscle and bone structure, as required in their pictorial practices.

In spatial representation, the invention of perspective, for example, plays a relevant role as a source of knowledge and rational understanding, using precise mathematical measurements to calculate height and width, to achieve symmetry, proportion and harmony. Some artists developed techniques or instruments derived from geometry and optics to approach their natural model with maximum accuracy. For example, instruments developed in the fifteenth and sixteenth centuries, called "drawing machines" by Albert Dürer (1471–1528) with "*el potrillo*", or Jacopo Barozzi Da Vignola (1507–1573) with the invention of the perspectograph (Sawday 2007). The accuracy thus achieved built a new paradigm that is still valid today, questioning the purely empirical sciences which used fragmentary deductive forms, to become a formal science dependent on logic and mathematics, and precise instruments (Foucault 1966) (Fig. 3.12).

Pioneer in naturalistic drawings are the sketches published in 1705 by Maria Sibylla Merian (1670 - 1717) under the title *Metamorphosis insectorum Surinamensium* (Fig. 3.13), illustrating the metamorphosis of insects. These sketches achieve in a single image multiple moments of its life cycle.

During the nineteenth century, impressive advances were made to describe the microscopic structure of biological tissues. Microscopic illustrations always imply a process of construction that includes the preparation of the sample, dying, and a set of illustration decisions regarding which structures to highlight in complex, tridimensional tissues (Figs. 3.14 and 3.15).

The appearance of photography gave rise to the scientific record and became a standard of proof *par excellence*. The first photographic picture used with scientific intent was by Charles Darwin in *The Expression of the Emotions in Humans and Animals* in 1872 (Prodger 1998). Later, in 1880, Alphonse Bertillon used photography to recreate an anthropometric identification system, the Bertillon method (Fig. 3.16).

At the beginning of the twentieth century, Ernst Haeckel, considered one of the greatest naturists of his time, published some of his more remarkable prints in the work titled *Kunstformen der Natur* (1904). With his plates, Haeckel has become an icon of artist skill at service of nature representation (Fig. 3.16).

As a more contemporary example, the mathematical studies of M. C. Escher (1898–1972), are magnificently expressed in his drawings, in which tiling, architecture and natural motives are employed. In 1959, the joint work of John Kendrew, Max Ferdinand Perutz and the artist-illustrator Irvin Geis made it possible to find the three-dimensional structure of the myoglobin molecule from two-dimensional data obtained by X-ray diffraction. This biological molecule, comprised by two thousand six hundred atoms, was represented three-dimensionally, even though no one has actually seen it, thanks to the remarkable spatial representation skills of Geis (Kemp 1998) (Fig. 3.17). This work earned Perutz and Kendrew the 1962 Nobel Prize in Chemistry.

One of the most popular scientifically driven visualisations is the representation of extinct beings based on paleontological evidence and modelling. When confronting fossilised remains, people may question the real appearance of the living creature, their colour or texture. Paleoartists,

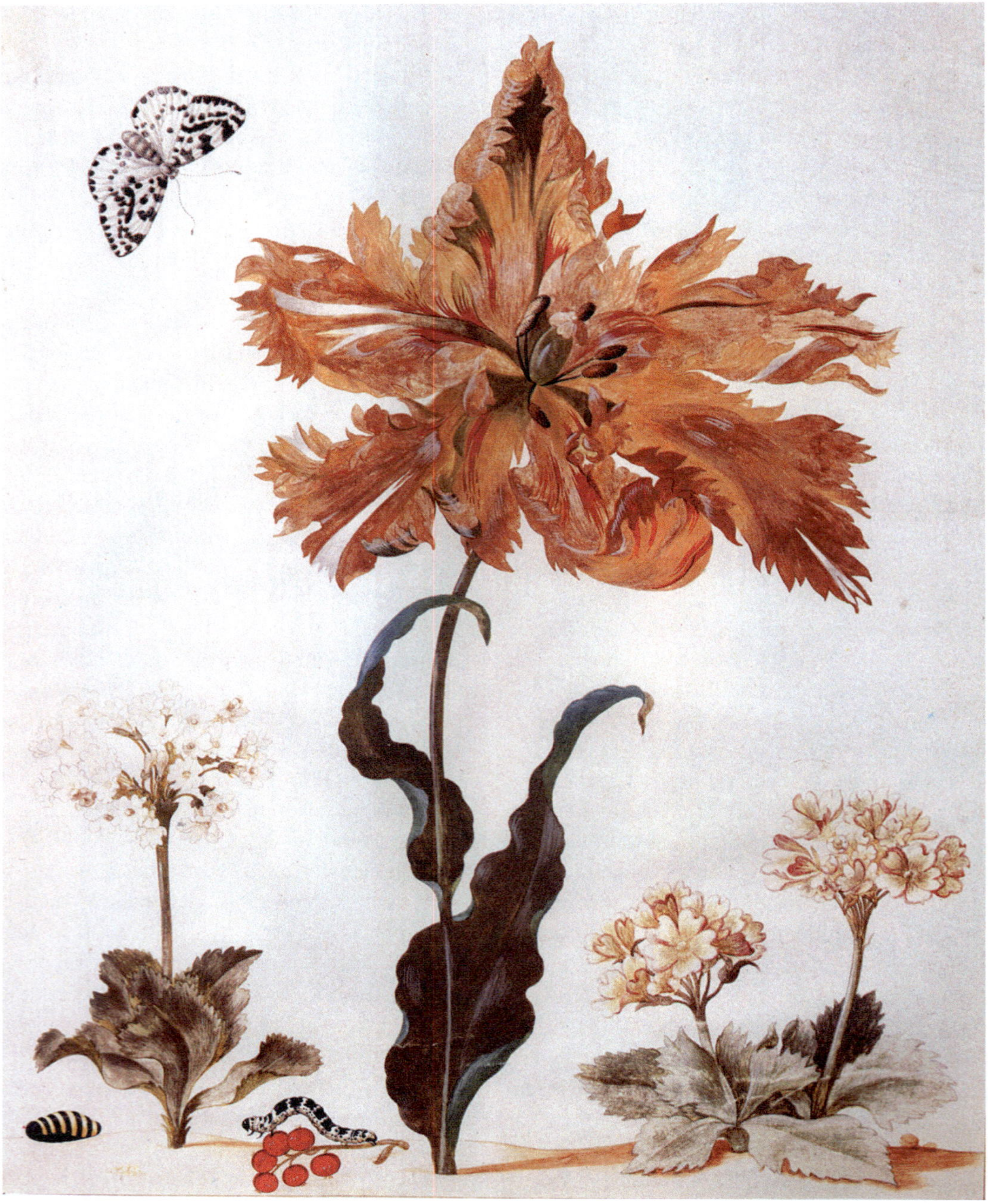

Fig. 3.12 Maria Sibylla Merian. 1705. *A Parrot Tulip, Auriculas, and Red Currants, with a Magpie Moth, its Caterpillar and Pupa*. https://es.wikipedia.org/wiki/Archivo:Maria-Sibylla-Merian-001.jpg. Reviewed January 2023. Public Domain

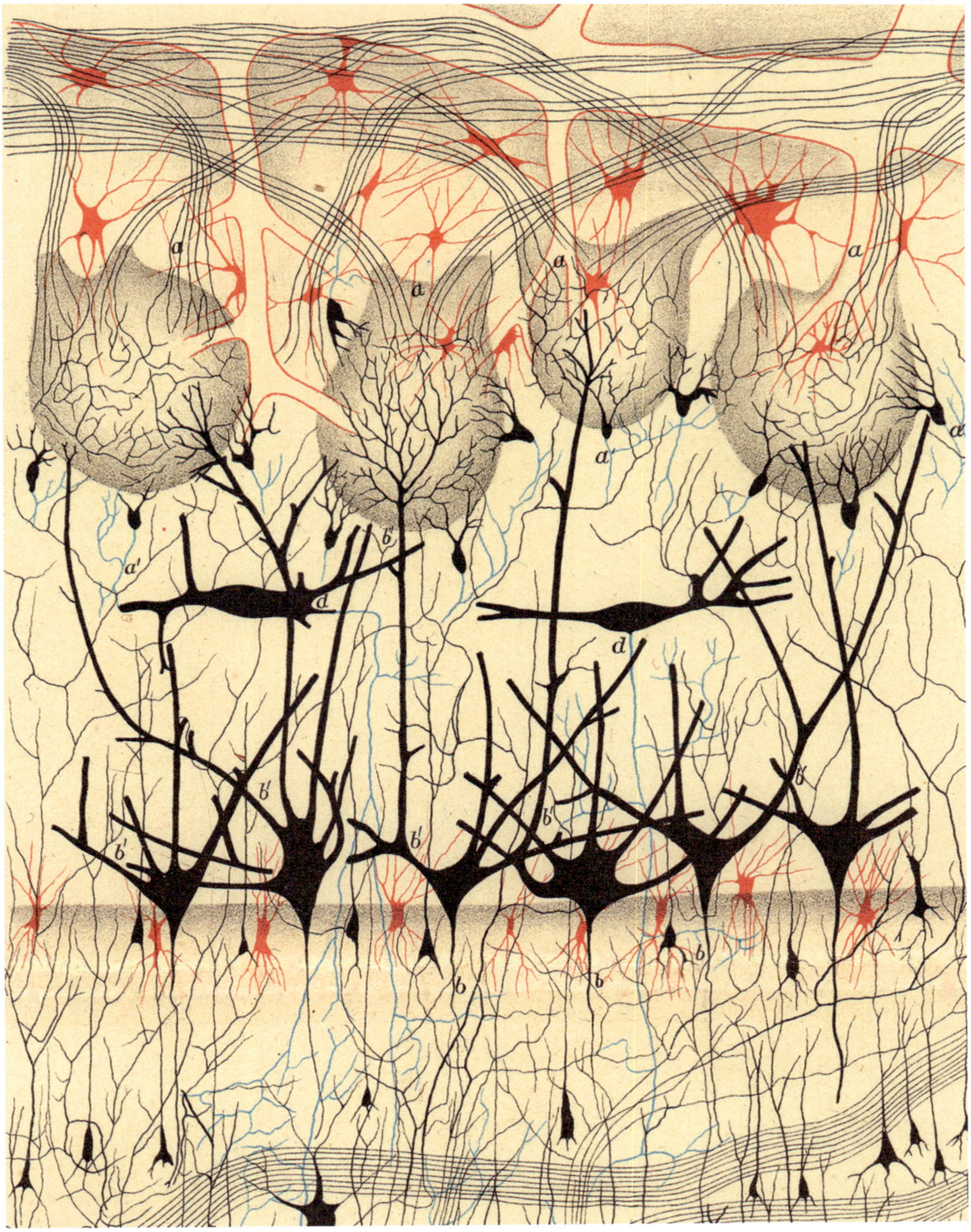

Fig. 3.13 Camillo Golgi. 1885. *Nerve cells. Sulla fina anatomia degli organi centrali del sistema nervoso. Studi. Reggio-Emilia, tipografia di Stefano Calderini e figlio*. Image courtesy of the Hagströmer Medico-Historical Library, Karolinska Institutet. Public domain

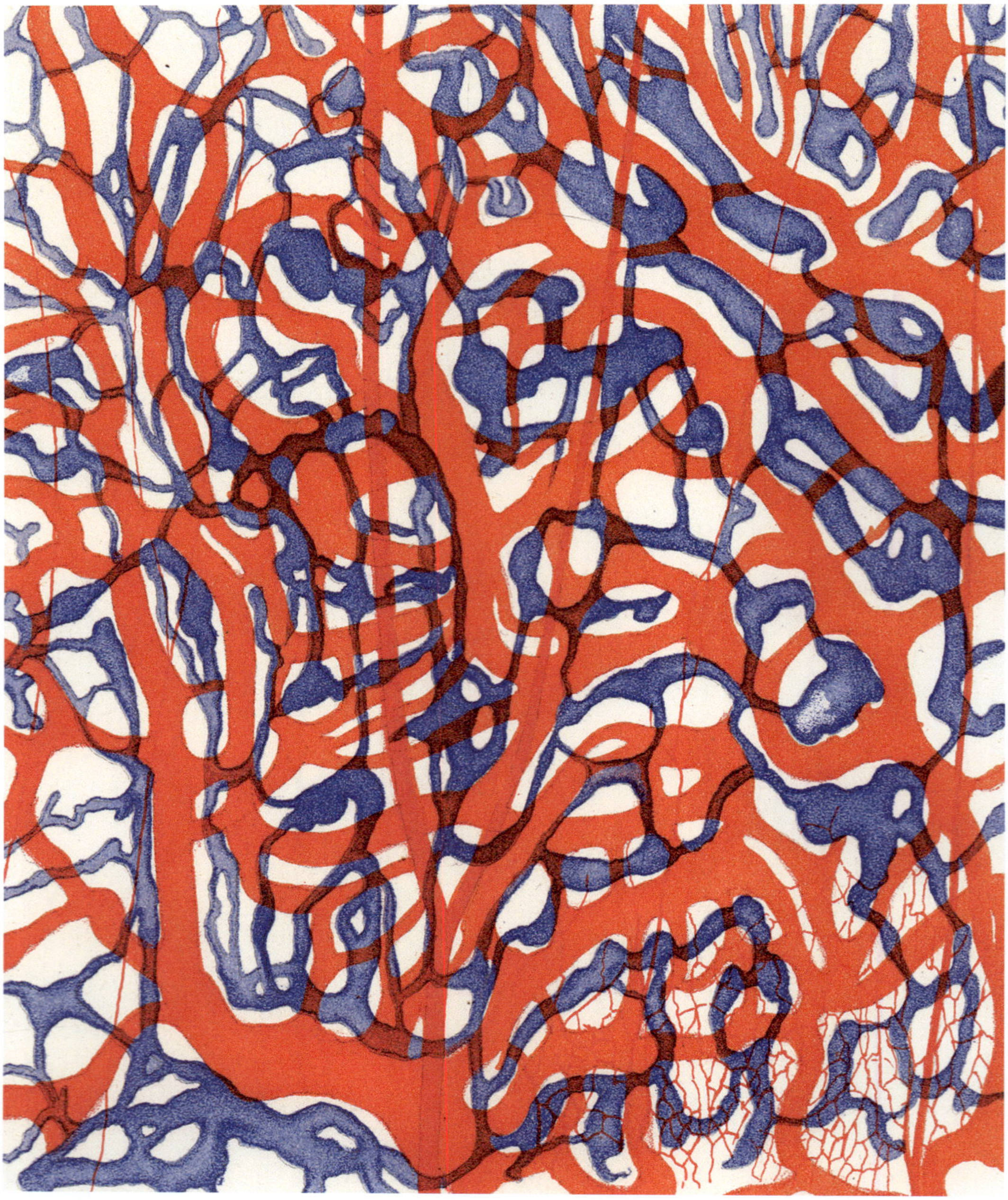

Fig. 3.14 Gustaf Retzius and Axel Key. 1875–76. *Part of olfactory mucosa from rabbit*. Studien in der Anatomie des Nervensystems und das Bindegewebes. Image courtesy of the Hagströmer Medico-Historical Library, Karolinska Institutet. https://www.alvin-portal.org/alvin/view.jsf?dswid=3457 Public domain

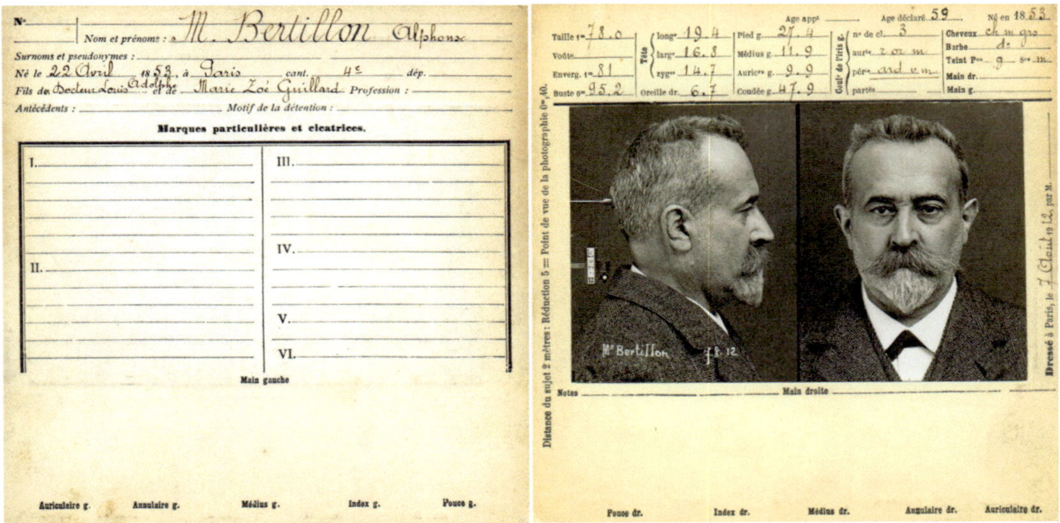

Fig. 3.15 Alphonse Bertillon. 1912. *Anthropometric data sheet (both sides) of Alphonse Bertillon*. Archives of Service Regional of Judicial Identity, Prefecture of Police. Paris. https://en.wikipedia.org/wiki/File:Bertillon,_Alphonse,_fiche_anthropom%C3%A9trique_recto-verso.jpg. Reviewed January 2023. Public Domain

the scientific illustrators who reconstruct prehistory in paintings, drawings, and sculptures, translate paleontological knowledge into plausible visual creations through sophisticated artistic skills and hundreds of fossil discoveries.

In Fig. 3.18, paleoartist Mauricio Alvarez, who lives and works on the island of Chiloe in Southern Chile, depicts the appearance and context of a new species of Ankylosaurus, *Stegouros elengassen*, based on paleontological findings and reconstructions.

The graphical representation of phenomena accessible by means of increasingly complicated digital visualisation technologies, far from delivering indisputable results, forces the researcher to make countless decisions about the image that "best" represents the phenomenon to be described. In fact, software applications specialised in processing different traditional analogical media (photography, video, audio) increase the alternatives and possibilities of data processing, making the influence of these applications even more significant to the final result. There is a return to the "subjective" image, after the partial replacement of illustration by photography at the beginning of the twentieth century (Wildevuur 2009). This issue is analysed in studies called Data Visualisation or Digital ethnography.

The impact achieved with this type of images is appreciated by science and the public through photography competitions such as *The Biomedical Image Award*, *Science and Engineering Visualization Challenge*, *Visions of Science Photographic Awards* and *The Art of Neuroscience* (Wildevuur 2009). Figure 3.19 shows an image of the Organ of Corti, in the inner ear, obtained by the award winning electron microscopist David Furness.

In short, this category is the one in which science holds almost all the authority. In professional investigation—from the lab, the field or theoretical work—arguments, evidence indications of various kinds and proper statistics are presented. However, a graphical representation is often necessary or desirable to capture, synthesise and present results and conclusions in the best and most convincing way. The artist's sensibility, abilities and creativity serve scientific ends and authorship. It is, no doubt, an asymmetric relationship. In turn, communication platforms of pure scientific works are traditional ones: articles in journals, congresses, specialised seminars. In the rare case that the graphic work comes out to

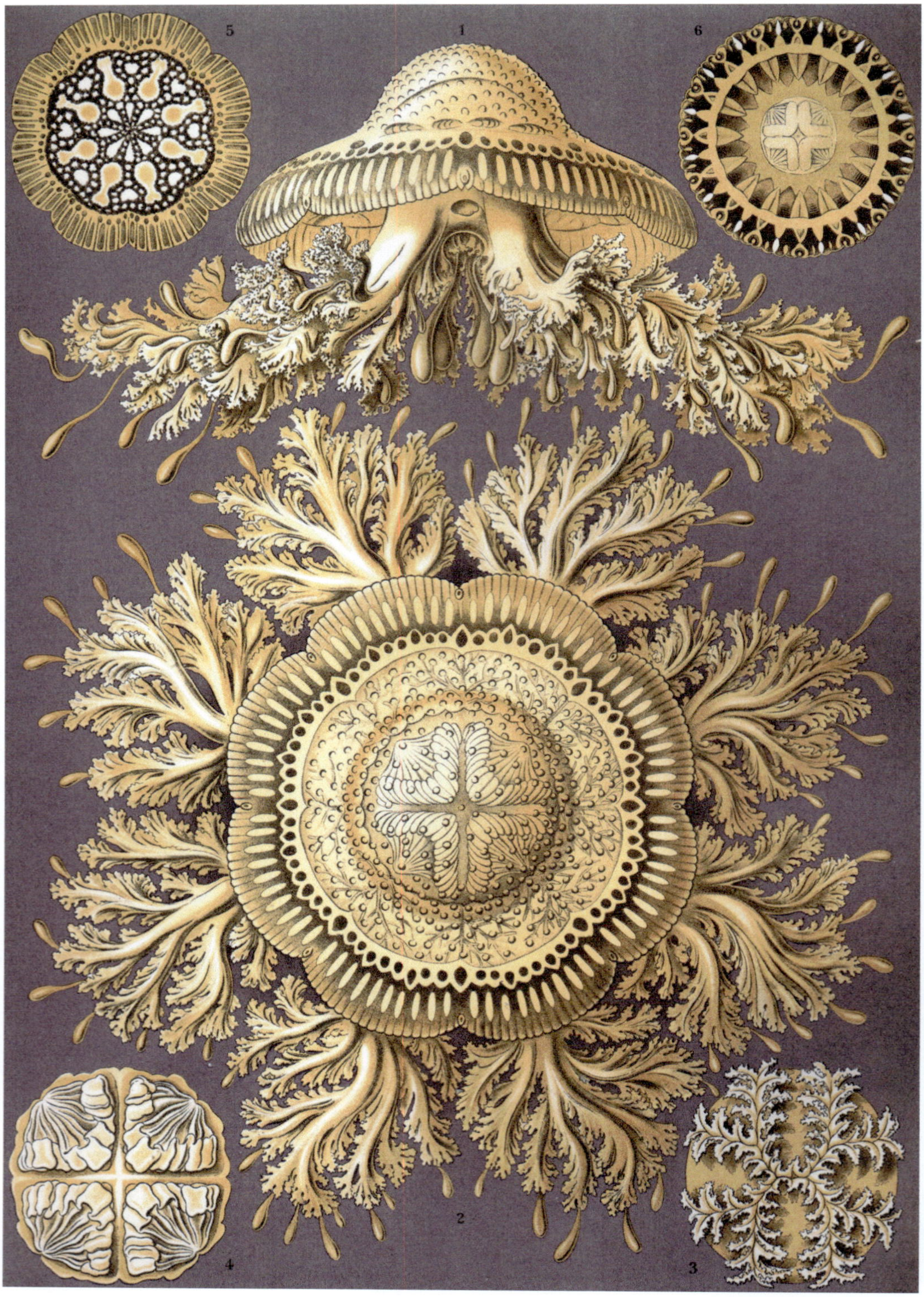

Fig. 3.16 Ernst Haeckel. 1904. *Discomedusae*. Plate 28 from the *Kunstformen der Natur.* This plate depicts anatomical features of *Reuma bellagemma* medusa. https://commons.wikimedia.org/wiki/File:Haeckel_Discomedusae_28.jpg. Reviewed August 2023. Public Domain

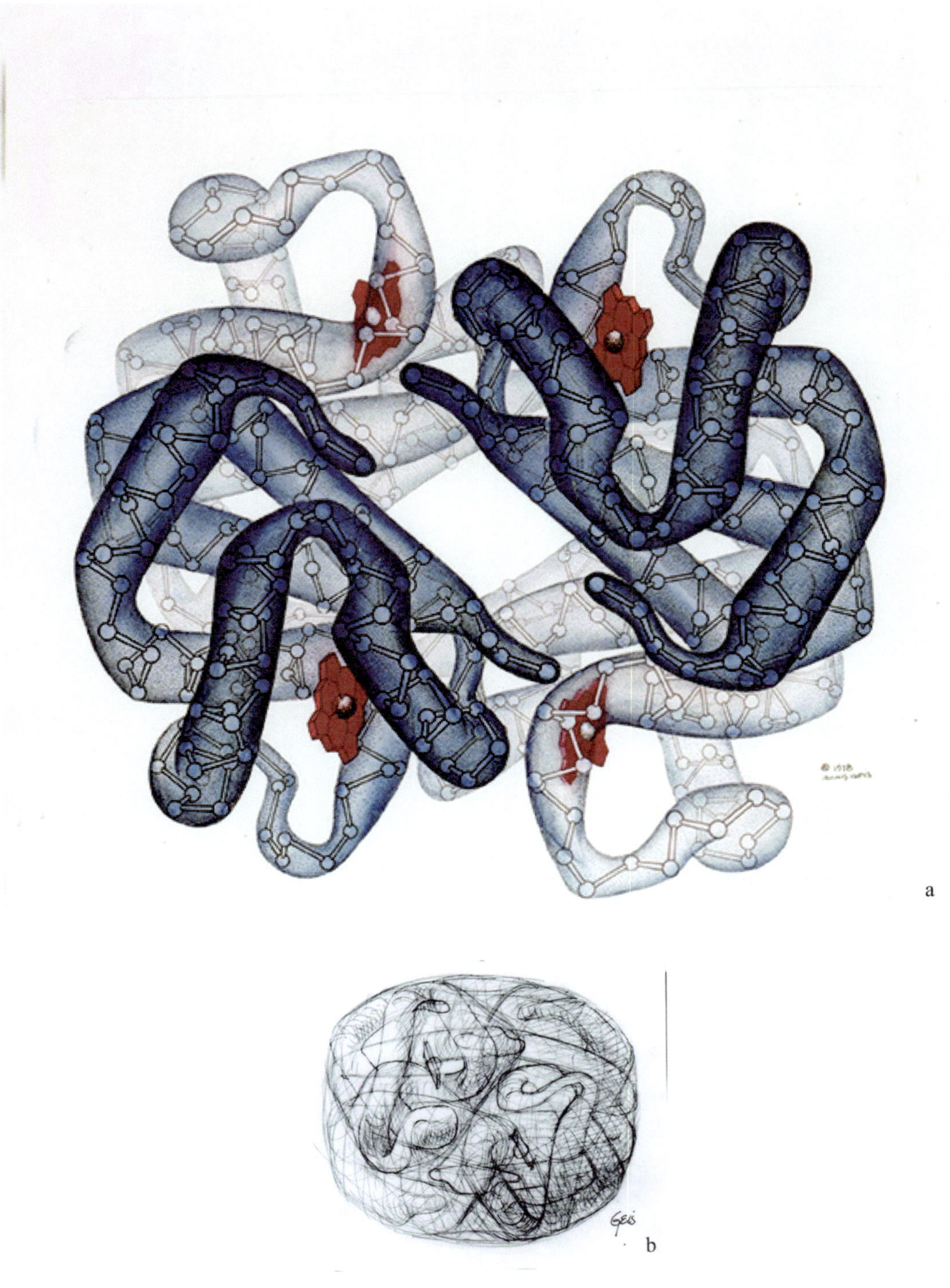

Fig. 3.17 **a** Irving Geis. (1978). *Oxy Hemoglobin Tetramer.* Illustration. **b** Irving Geis. (1984). *Hemoglobin - study of whole molecule as single form*. Illustration. Rights owned by Howard Hughes Medical Institute. Images Courtesy of Sandy Geis

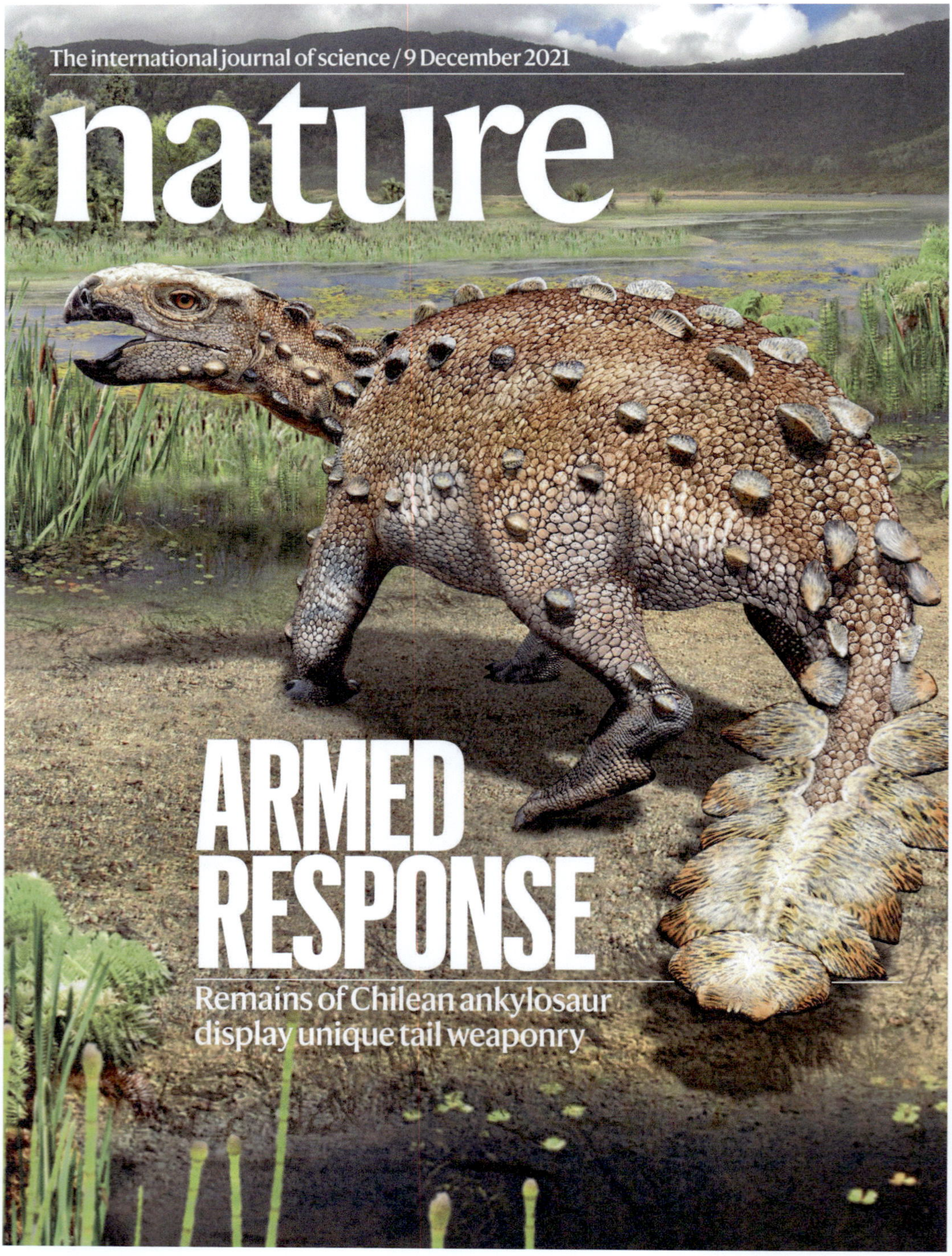

Fig. 3.18 Mauricio Alvarez. 2021. *Nature's cover with the new species of Ankylosaurus, Stegouros elengassen,* published in December 2021. 3d base of the illustration by Gabriel Díaz (@paleogdy) and photographic base by Fernando Borquez (@ferborq). The paleontological work was developed by the Red Paleonica of the University of Chile, The Instituto Antártico Chileno, and the Museo de Historia Natural of Chile. Image courtesy of the author

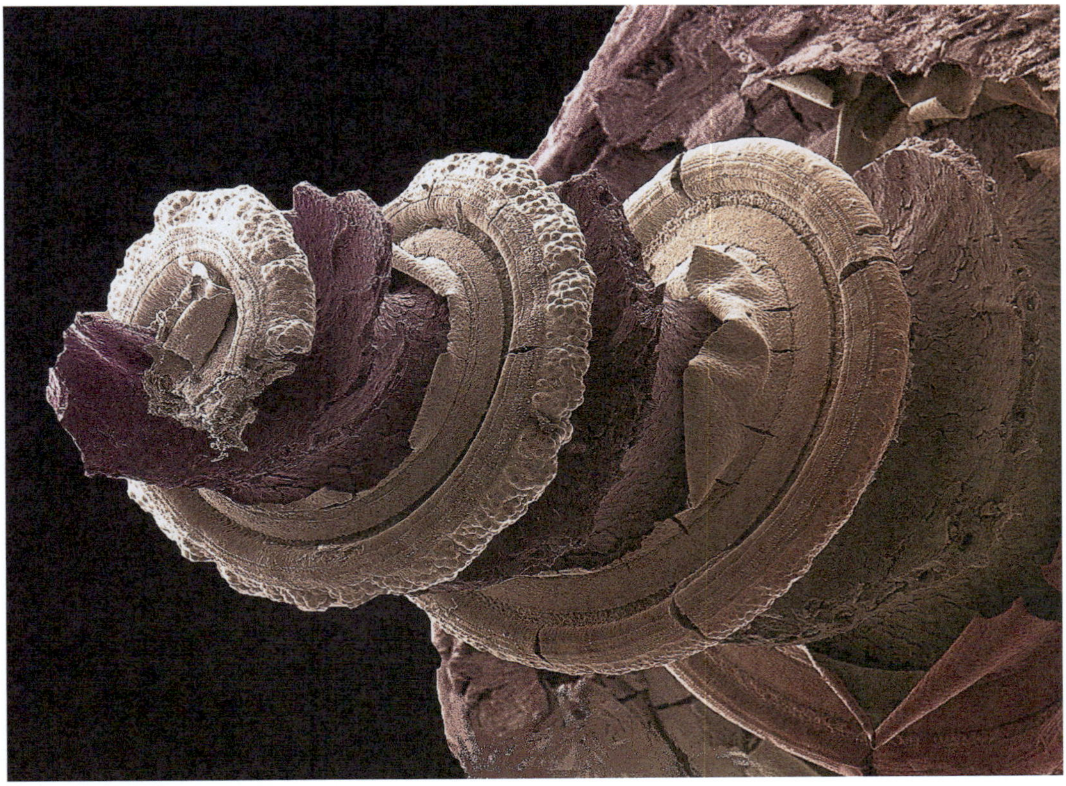

Fig. 3.19 David Furness. 1997. *Cochlea of the inner ear—coloured—brown*. A scanning electron micrograph of the inside of a guinea pig inner ear showing the hearing organ, or cochlea. Biomedical Image Awards 1997. Image courtesy of the author

the public in an open way, it would be a secondary consequence of the initial purpose.

c. **Works that use science as a model or reference for art**

   Artists may freely represent a natural phenomenon—previously described by some scientific discipline—using different strategies and techniques like graphics, sculpture, audio-visual or interactive installations. The artist generates visualisations of the subject, where aspects of its form, organisation or dynamics are recorded, expanded, synthesised or highlighted in an expressive form.

The piece *Out of Order* (2003) by artist Richard Deacon (Fig. 3.20) is an example of this type of work, created on the 50th anniversary of the publication of the structure of DNA. The artist develops a complex, large format work composed of loops of wood and steel twisted in space, which expresses what he sees as the essence of large organic molecules, without being a figurative representation of any particular one (Kemp 2003, 2005).

> Deacon's magnificent sculpture is no longer about DNA .... *Out of Order* (suggests), among other things, the emergence of unpredictable complexity from a limited set of basic parts... Its rootedness in a conceptual process akin to that which allowed Watson and Crick to see the structure of DNA remains central to its formation. (Kemp 2003, parr 8)

In 2003, the MOMA PPS1 in New York presented an exhibition called *Signatures of the Invisible*, the result of a collaboration between artists from the London Institute and particle physicists from the Conseil Européen pour la

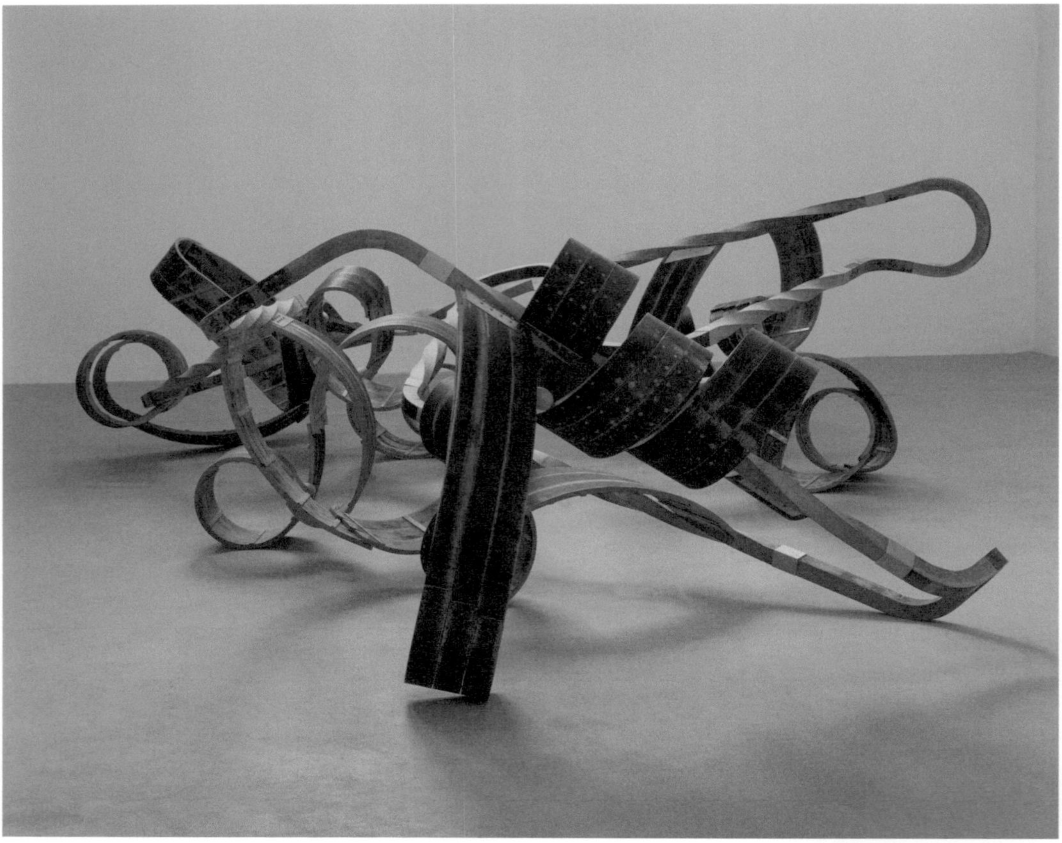

Fig. 3.20 Richard Deacon. 2003. *Out of order*. Wood (steamed oak), stainless steel. Photo by Dave Morgan

Recherche Nucléaire (CERN). This exhibition was the brainchild of artist Ken McMullen and CERN Theoretical Division Director Maurice Jacob. Twelve renowned artists were given free access to CERN to discuss the enigmatic world of high-energy physics with experimental and theoretical researchers. The project was motivated by the conviction that artistic practice requires a commitment and negotiation between the artist's subjective or intuitive processes and experimental access to the material world (McMullen 2005). The works produced are not illustrative of scientific results from CERN, but of metaphorical operations derived from the encounter between artists and scientists. Among the participating artists were Roger Ackling (Great Britain), who worked with the magnification and focus of sunlight on wood; Mel Chin (United States), who during the three months of the exhibition cultivated certain plants that allowed him to extract heavy metals from contaminated soil, with which he then built a pencil and drew with it; and a physicist from CERN, Jérôme Basserode (France), who worked on the concept of time and its limits using large metal spinning tops.

Another example is *Floralis Genérica* (Fig. 3.21), a spectacular metallic sculpture erected in 2002 in the United Nations Plaza in the City of Buenos Aires, Argentina. The piece, created by Argentine architect Eduardo

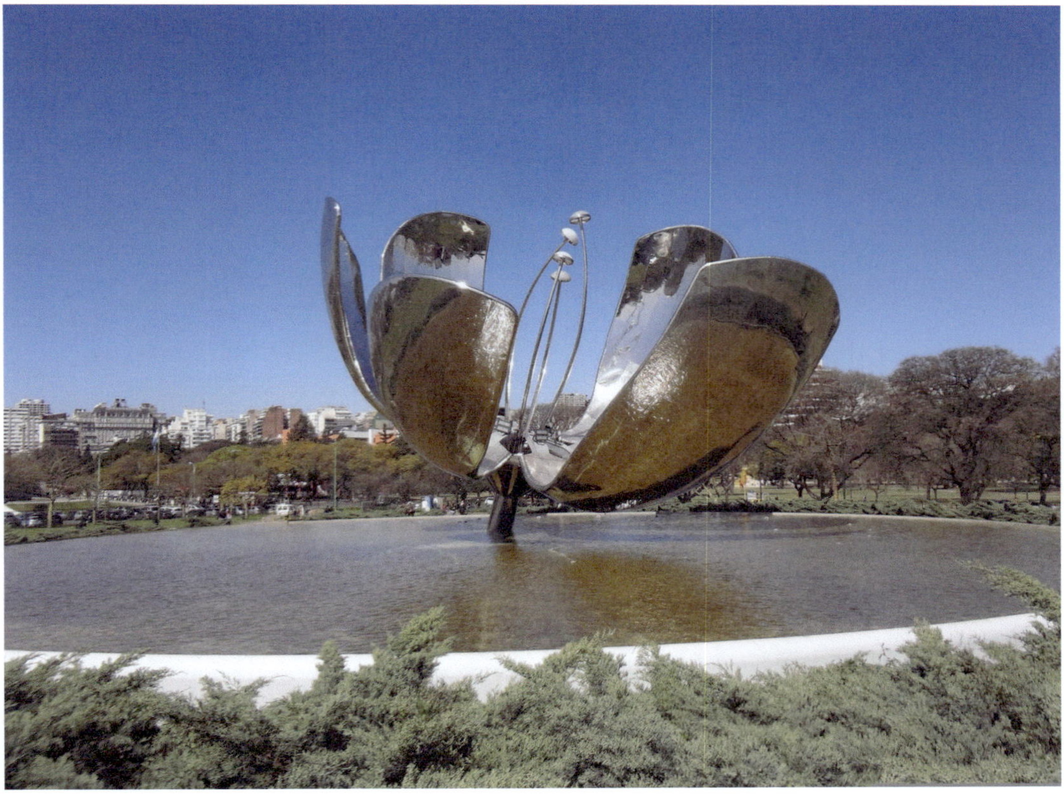

Fig. 3.21 Eduardo Catalano. 2002. *Floris Generica*. https://pixabay.com/es/photos/floralis-gen%C3%A9rico-buenos-aires-949177/?download. Image courtesy of Alexandria account – ID 139,904 2013 download from Pixabay. Download February 5, 2023

Catalano, is a flower that opens during the day and closes at night, emulating the process of photonasty. In a similar line, at the Science Center Technorama in Winterthur, Switzerland (1969), works such as *Tornado* (1990) by Ned Kahn and *Magic Wave* (2008) by Reuben Margolin stand out as installative representations of natural phenomena.

The work *Dripstone Machine* (1996–2496), by Bogomir Ecker (Fig. 3.22) is an installation/experiment conducted at the Hamburger Kunsthalle in 1996 that will be completed in approximately five hundred years. The artwork consists of collecting rainwater from a roof gutter, which passes through carbon dioxide and calcium carbonate before dripping onto a marble slab. The objective of the work is to achieve the formation of a five-centimetre-high stalagmite on the marble. Its goal has a scientific basis, but no one knows for sure what will happen in this "experimental" approach, juxtaposing an explanation with the reality of the phenomenon.

Antoni Rayzhekov's work, *Music of bacteria colonies grown from money* (2018) (Fig. 3.23) critically examines the notion of value, moving it to the realm of the biological and then to that of sound, to suggest a novel perspective towards the distribution and circulation of money. It proposes a notion of "sonic" value, converting the value of the currency depending on its exponential growth (the growth of bacteria).

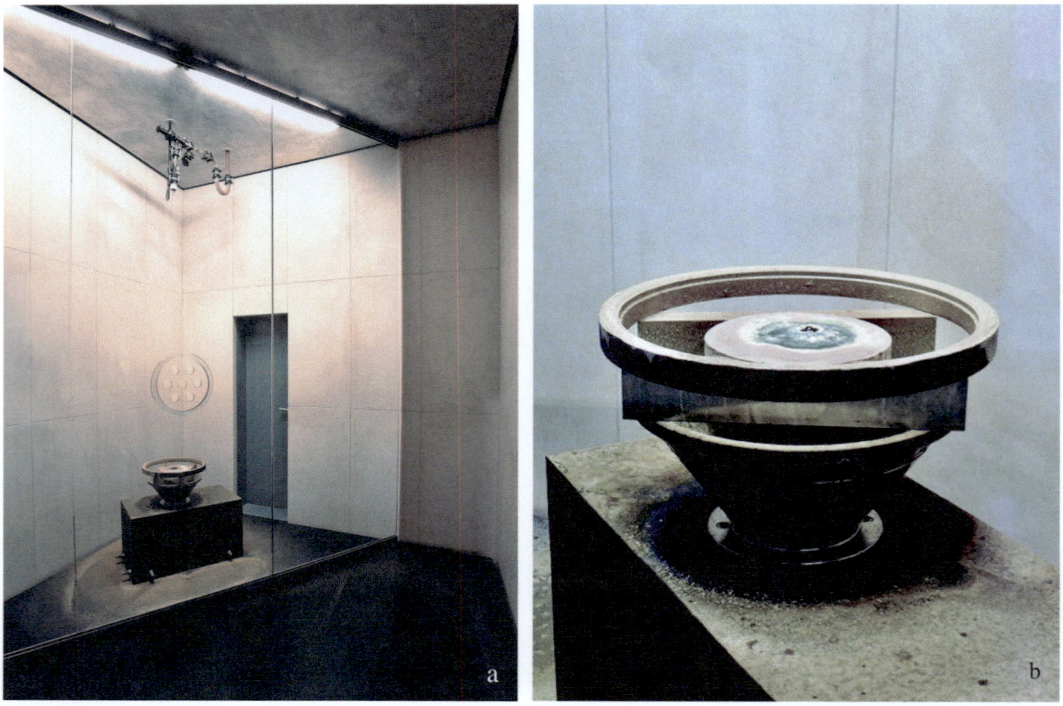

Fig. 3.22 Bogomir Ecker. 1996–2496. *Dripstone Machine.* Room installation. **a** General view. **b** Detail. Image courtesy of the author

In this category, a more symmetrical relation of authority can be identified in the art-science integration. Even though the "model" or reference may come from science, the purpose is artistic, and the proposal is a free expressive/declarative act by the artist. Accordingly, the media platforms resorted to are those of art galleries, museums, public spaces, art journals, and any mass media.

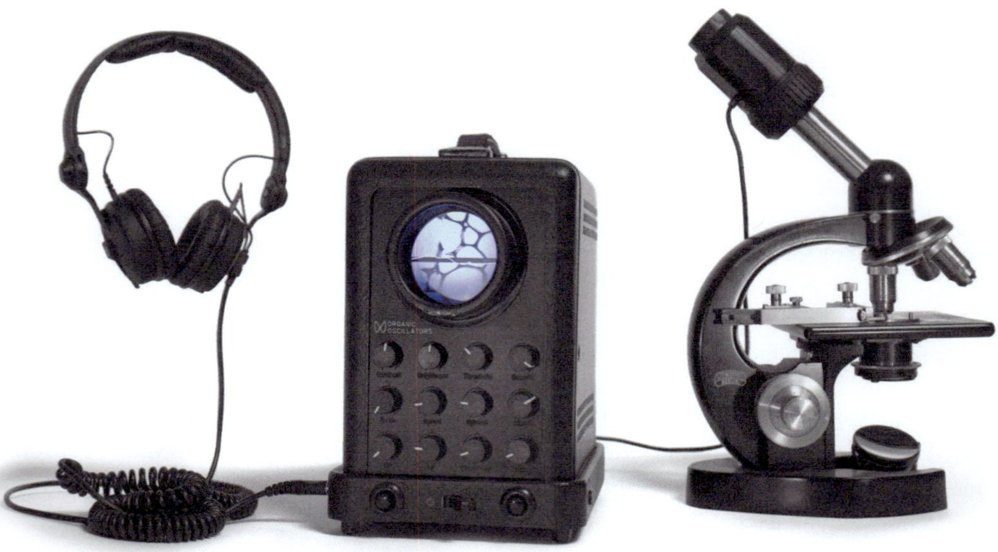

Fig. 3.23 Antoni Rayzhekov. 2018. *Music of bacteria colonies grown from money*. Interactive audiovisual installation and performance. Image courtesy of the author

d. Works in which science is explored, problematized or questioned by art

According to the German philosopher Immanuel Kant, human beings have two types of faculties. The first is understanding, which is based on certain subjective sensations, and the second is reason, which is the ability to judge something. Although Kant attributed this second critical faculty to philosophy, it is also developed in artistic production. There are many artists who work with a particular sense of curiosity and sensitivity which is activated or relates to a critical stance, political denunciation or activism. For some artists, political engagement is the only valid purpose of contemporary art.

In this line, science fiction is an interesting illustration of mutual influence and co-dependence of science and art. As works of fiction (novels, short stories and tales, comics, cinema, television series, graphics) this genre is open to scenarios that are fuzzily located between the possible and the impossible (Fig. 3.24). They move the limits of current theories and models as a result of speculative or prospective technological developments. Thus inspired, these works create plausible worlds, giving rise to imaginaries of the future that become, in many cases, ideals of progress, utopias and dystopias (for example, space exploration, biological manipulation, mastery of cybernetics and artificial intelligence). Science fiction becomes "…part of a strong trend towards employing speculative fiction explicitly for futurological purposes and towards taking science fiction seriously in mainstream, socially critical discussions" (Grillmayr 2020, p. 141).

These imaginaries and speculations can reinforce or question the public's interest in certain topics, and therefore influence the existence of large-scale research programs. Among these it can be mentioned the space programs to Mars, the Human Genome project (International Human Genome Sequencing Consortium 2001), or The

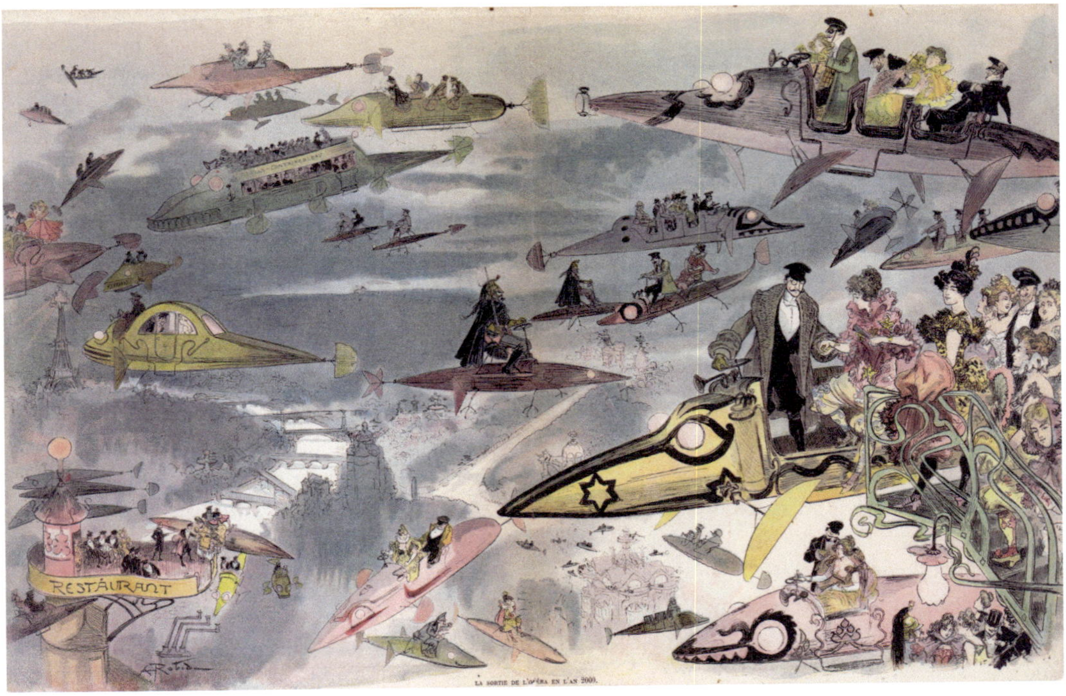

Fig. 3.24 Albert Robida. 1882. *A night at the opera in the year 2000*. Print. Library of the Congress United States America. https://commons.wikimedia.org/wiki/File:A_night_at_the_opera_in_the_year_2000,_cartoon_by_Albert_Robida,_ca._1882.jpg. Reviewed February, 2023. Public Domain

Decade of the Brain, an effort aimed to achieve a qualitative leap in the understanding of the brain (see [4]), all recurrent themes of science fiction.

In a more instrumental and experimental approach, the use of biological material could be identified as the scientific discipline most resorted to by artists and gives rise to Bioart (Kac 2007; Mitchell 2010). Bioart is an artistic manifestation with multiple layers of significance and implying different subtypes of biological areas. It arose as a response to the impact of scientific discoveries in biology during the twentieth and twenty-first centuries, and today has an association based in Helsinki, the Bioart Society.

An early reference of biology and art is the work by artist Luis Fernando Benedit (Argentina 1937–2011), the *Biotrón.* Presented in the 35th Venice Biennale in 1970, was a collaborative effort with ethologist Josué Núñez and experimental psychologist Antonio M. Battro. The *Biotron* explored human-animal interactions with an installation placed inside the expositive hall (Badagliacca 2021). Another pioneer in this field is artist Joe Davis, staff member of MIT, Harvard Medical School and the University of Kentucky. Thirteen years ago Joe Davis, and a group of geneticists from Harvard, encoded the first graphic image in the DNA of a bacteria (Davis 1996).

Performance artists *Stelar* and *Orlan* subject their bodies to modifications and extensions. They challenge conceptions of the body and medicine and humanity's relationship with technology and the natural environment (Macneill and Ferran 2011; Ede 2005). Bioartists are concerned about the risks related to biotechnological developments but also enthusiastic about the possibilities raised by the gradual openness of research labs, with cheaper and easier experimental technology.

The famous Brazilian bioartist Eduardo Kac raised controversial issues with his notion of transgenic art in 2000. In his piece *Genesis* (1998/1999), he questioned the use of technology and its dialogical relationship with biology and the genetic, linking the Judeo-Christian tradition of the creation of the world with the manipulation of genetic material. Art as a public policy medium. Another work by Kac is *GFP Bunny* (2000), consisting of the genetic modification of a rabbit so that it expresses fluorescent green coloration when illuminated with blue light in the dark. This genetically modified rabbit now becomes an easy prey in the wild (Kac 2007). Or Portuguese artist Marta de Meneses with her seminal bio-artistic work and her current prolific participation in residences and exhibitions.[5] In *Anti-Marta* (2018) she explores and questions the meaning of self and otherness as a poetic operation through an immunological exercise in which she and her husband graft and exchange a piece of each other's skin to experience the effect of immunological rejection (For a review of bioart and an extensive bibliography[6]).

Tackling environmental issues, artist Anne Duk Hee Jordan's video installation *Ziggy and the Starfish* (2016–2018) examines the effect of pollution and climate change in marine life, showing how higher temperatures are changing the reproduction and sexuality of the ocean's inhabitants. Amy Balkin's work, *Today's CO_2 Spot Price* (2009) is a real-time installation of climate market data for the European Union from July to September 2009. The daily spot price for CO_2 emission permits was updated every day over the duration of the exhibition (Fig. 3.25) from the European Energy Exchange. Her project *Public Smog* (2004–) includes an effort to have the Earth's atmosphere included on UNESCO's World Heritage List. By buying carbon credits and making them inaccessible to markets, the work ironizes the fact that the atmosphere itself is indivisible; pollution transcends geographical, financial and political boundaries and, as such, the idea of a clean air park is unrealistic.

From a different scientific field, the work of Kit Galloway and Sherrie Rabinowitz, *Satellite Arts Project: A Space With No Geographical Boundaries* (1977), was made in collaboration with NASA, and was the first interactive dance

[4] For a Decade of the brain see the website of the Congress of the United States of America: https://www.loc.gov/loc/brain/. Reviewed, February 6, 2023.

[5] For more information, see her website https://cultivamoscultura.com, Reviewed on March 19, 2023.

[6] See Yetisen et al. (2015).

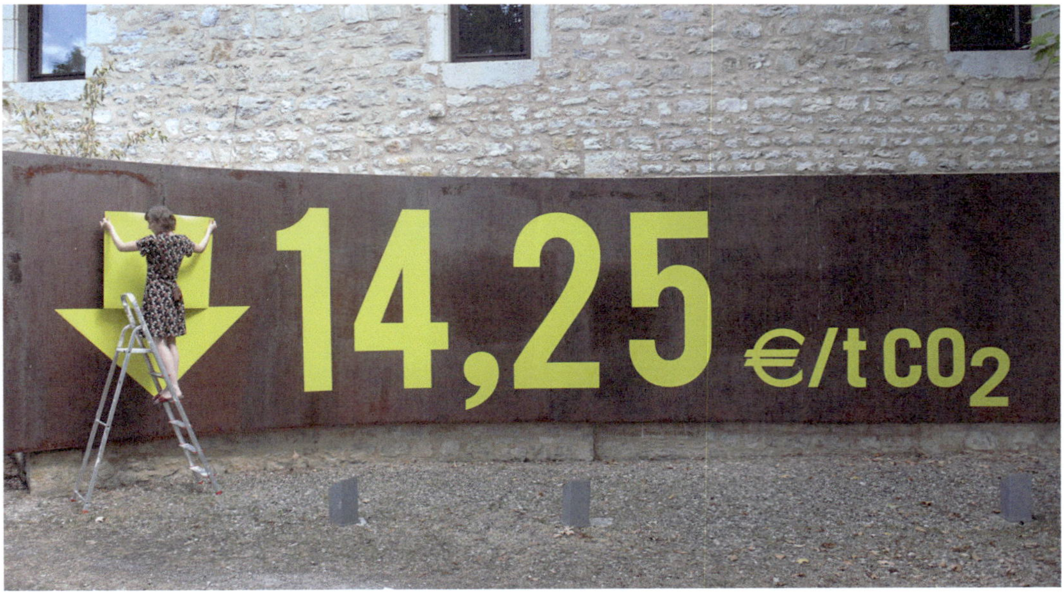

Fig. 3.25 Amy Balkin. 2009. *Today's CO_2 Spot Price*. Custom-cut magnetic numbering system. Thanks to Martine Michard, Hélène Maury, Stephane Marchais, and Monika Kulicka. Image courtesy of the author

performance communicated by satellite. Using the US-Canadian CTS satellite, dancers located 5000 km apart in Maryland and California, were electronically composited into a single image on display on monitors at each location, creating a "geographically boundless space" in which the live performance unfolds. This work questions whether human relations that are inherently social in nature require proximity, and in so doing, anticipates what actually occurred later with the explosion of the Internet and our move to social interaction at a distance.

At the end of the 1970s, artist Juan Downey developed a pioneering work living with the Yanomami tribe in the Venezuelan Amazon (Fig. 3.26). He inverted Western anthropological forms of registration, handing over the video camera to the Yanomami, dismantling the interpretation that the observer can obtain from his object of study (Guagnini 2013).

Finally, with the massification of digital networks, a new generation began to work with different technological devices and to generate networks and activities closer to scientific experimental systems, but with the additional element of political and cultural dissent. Examples of this type are The *Chaos Computer Club* founded in Hamburg in 1981 and the emergence of the *neo-avant-garde* of net.art in Europe and the United States.

In this fourth category, even though the subject behind the work is scientific, the issue to be explored or problematized is freely chosen by the artist and is treated by artistic means. The scientific subject can vary widely while remaining functional to the question of the artist. Thus, this category demonstrates an asymmetric integration between science and an artistic authority where the expositive formats and media employed are artistic. Nevertheless, some of these works have been so professionally developed and well documented that their results have reached specialised scientific platforms, for example, the work of Juan Downey, Joe Davis, or Oron Catts and Ionat Zurr.

e. **Works that provide a scientific explanation of the aesthetic experience**

Artistic practice is a distinctive feature of humans. Its emergence is a key milestone in our evolution, and it is manifested in all cultures. A methodical approach to the phenomenon of art can be traced to a time long

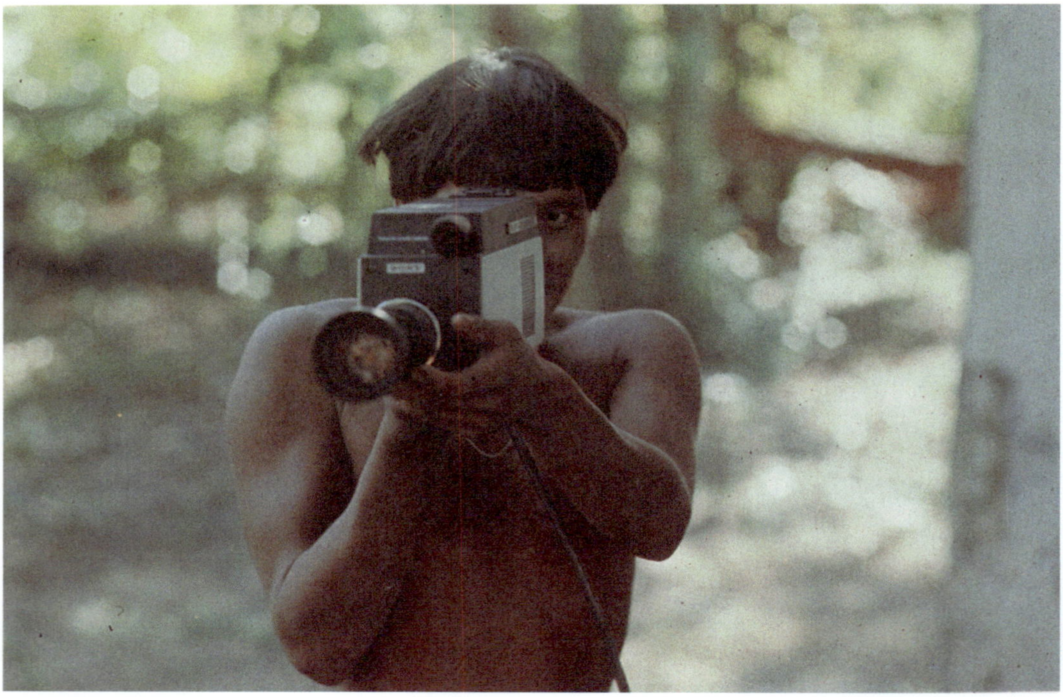

Fig. 3.26 Juan Downey. 1979. Still from the video *The Laughing Alligator*. Image courtesy of Engel Foundation. Santiago, Chile

> before the origin of the scientific method, with profound philosophical implications. Science has pondered the origin of perceptual sensations and of our sense of beauty, and how they relate to each other.

From the tradition that links art with beauty, a theory, model, image, scientific result, or the mere encounter with forms of nature can be valued as an aesthetic experience and equated to a work of art. Reciprocally, for Lavin, when the history of art is told, a "natural science of the human spirit" is created (1994, p.18) just as mathematics would be a natural science of the intellect. In a similar vein, Feyerabend (1994) proposes that art is the result of nature insofar as it is produced by human beings who are part of nature.

Aesthetics, the study of the principles of art and beauty, was addressed in the nineteenth century by the psychology of aesthetics, a newly founded discipline in search of a general unified theory that would describe and explain the mental factors involved in the aesthetic experience. The philosopher and psychologist Gustav Theodor Fechner published the *Vorschule der Aesthetik* (School of Aesthetics) in 1876, establishing Experimental Aesthetics, the second oldest branch of experimental psychology after psychophysics. Fechner proposes the concept of *aesthetic threshold* as those stimuli that trigger experiences as aesthetically significant, or pleasant and unpleasant (Jacobsen 2006). He also seeks the objective bases underlying "the evaluation or production of beauty, ugliness, loveliness, harmony, elegance, proportion, or charm" (p. 155). In his article, Jacobsen lists a series of variables that can influence aesthetic experience and judgement: symmetry, complexity, simplicity, novelty, proportion, composition, semantic content, as well as the emotional state of the person, the degree of interest in the stimulus, the search for social acceptance, the educational, historical and cultural background of the observer, and aspects of the context, such as the place and time of exposure to the aesthetic stimulus.

Also during the nineteenth century, Theodore Lipps (1851–1914) linked the aesthetic experience to the concept of empathy, understood as the experience, in oneself, of the feelings and emotions of the other as a basic form of connection between two subjects (Hunsdahl 1966). This emotional projection allows the immediate comprehension of the expression of the other (Lipps 1923, as cited in Mongarde 2000) even before its intellectual grasp.

As a biological bases of empathy and art, neurophysiologist Giacomo Rizzolatti and his team at the university of Parma discovered, in 1996, a particular type of neuron found in premotor, motor, somatosensory and body-image areas of the brain's cortex: the *mirror neurons*. These neurons discharge when an action is observed in others and when the action is executed by the subject. This neural response may account for empathetic feelings of involvement, enabling an experiential understanding of objects and body movements in works of art (Freedberg and Gallese 2007). For these authors, empathy may be understood as an *embodied simulation*, that is, a "mechanism through which the actions, emotions or sensations we see, activate our own internal representations of the body" (p. 198). Moreover, through this embodiment process, two aspects of the aesthetic experience may be approached: first, the representational content of the artwork (actions, intentions, objects, emotions, etc.) and second, the empathetic feelings embodied in the observer of the visible traces implied in the creation of the piece, for example the tactile texture of the piece, vigorous modelling in clay or painting, brushstrokes, drippings, and signs of the movement of the hand more generally (op. cit). In relation to Fig. 3.27, these authors explain that these experiences "also applies to the cut

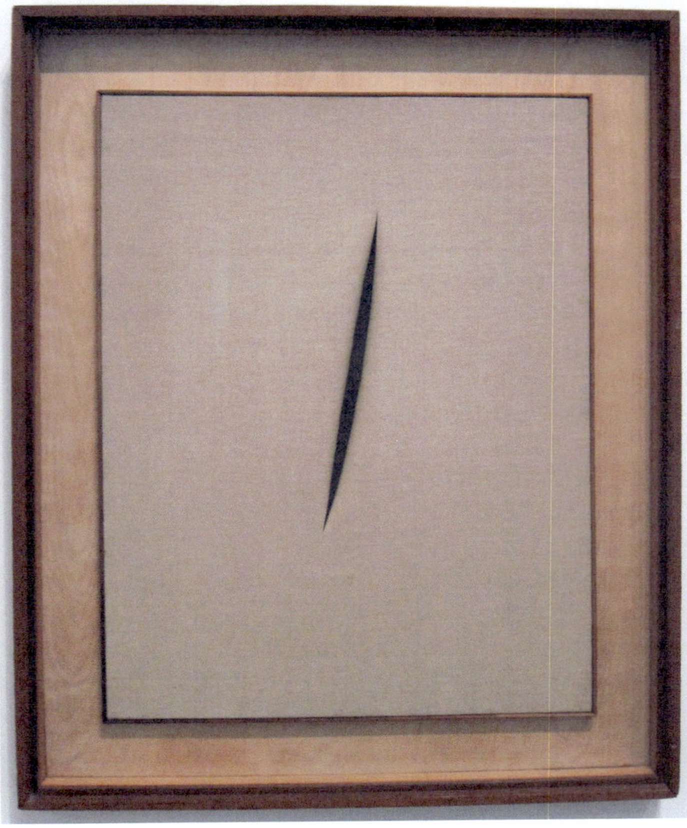

Fig. 3.27 Lucio Fontana. 1960. *Spatial Concept "Waiting"*. Photo Wmpearl. https://commons.wikimedia.org/wiki/File:Spatial_Concept_%27Waiting%27,_cut_canvas_by_Lucio_Fontana,_Tate_Modern.JPG. Reviewed June, 2023. Public Domain

canvases of Lucio Fontana where sight of the slashed canvas invites a sense of empathetic movement that seems to coincide with the gesture felt to have produced the tear"(p. 197).

Figurative and abstract images have been used in psychology to gain insight into subjective emotional states. The Rorschach Test, developed by Hermann Rorschach in 1921, consists of a series of indefinite images whose interpretation is used as a diagnostic tool. In the same line, the Rosenzweig Test is a set of twenty-four cartoon-like images developed in 1948, that relay frustration to aggression. The enigmatic drawings created by Gustav Jung in relation to the unconscious is another remarkable example.

The Nobel Prize-winning molecular biologist, neuroscientist and art lover Eric R. Kandel believes that science can help to understand how we perceive, appreciate and comprehend works of art through the senses, the body and the brain (2016). Initially, the neurophysiology of perception searched for the biological mechanisms behind what was conceived as a passive grasp of the world around us, assuming that a perceptual system is an arrangement of sensors that capture information from the outside or external world. This information was assumed to be processed in the brain to produce a bodily response. However, the physiology of perception has found active and self-referential neural mechanisms that construct the world, coherent with the historical survival of the species and its individual development, rather than the capture of a pre-existing objective, exterior reality that determine by itself the behaviour of the subject (Maturana and Mpodozis 1987).

Additionally, a recurrent starting point in biological approaches to art is that aesthetic experiences are the result of "adaptive advantages" that contribute to the "survival of the fittest", with the corresponding perpetuation of certain genes. This current of thought, although highly questionable, is present explicitly or implicitly in many of the approaches that seek to explain art as a biological phenomenon. That is to say, the neurobiology of art or neuroaesthetics entails, together with several indisputable experimental results, debatable conceptions of both what art is and what we are as human beings. In the article *The Science of Art*, Vilayanur S. Ramachandran and philosopher William Hirstein (1999) present a list of ten universal perceptual laws, underlying the experience of aesthetic pleasure.[7] This list assumes that the perceptual pleasure or sensation produced by these universal laws is justified in terms of adaptive advantages.

With a clear physiological approach, Tomohiro Ishizu and Semir Zeki's article *Toward A Brain-Based Theory of Beauty* (2011) uses functional magnetic resonance imaging (fMRI) to record the activity of one cortical area located in the medial orbito-frontal cortex, that is "active during (both) the experience of musical beauty and visual beauty, ... overlapping almost completely within it" (2011 p. 1) (Fig. 3.28). Using a similar protocol, they later discover that brain activity related to mathematical beauty is also located in the medial orbito-frontal cortex (Zeki et al. 2014). These results may suggest the existence of a specific area in the brain activated by a general feeling or sense of beauty. To operationalise the idea of beauty, the authors resorted to participants' subjective report of "beautiful or not beautiful" in relation to a set of different motives or figures presented to them.

Another current experimental approach to aesthetics is taking place in the Laboratory for Cognitive Research and Art History (CReA Lab, University of Vienna), with their special interest in eye movements during the observation of artworks. For example, *The Last Supper* by Lucas Cranach the Elder (1547) is used in one study to track the movement of the eye as it travels through the different parts of the canvas, making it possible to identify how the trajectory and linear frequency of the observer's gaze constructs the global abstraction of the image (Rosenberg and Klein 2015; Sancarlo et al. 2020) (Fig. 3.29).

[7]The complete list of the ten universal perceptual laws in the brain is 1. The Peak Shift principle; 2. Perceptual grouping and binding; 3. Isolating a single module and allocating attention; 4. Contrast extraction; 5. Perceptual problem solving; 6. Symmetry; 7. The generic viewpoint and the Bayesian logic of perception; 8. Art as metaphor; 9. Repetition, rhythm and orderliness; 10. Balance. (Ramachandran and Hirstein 1999).

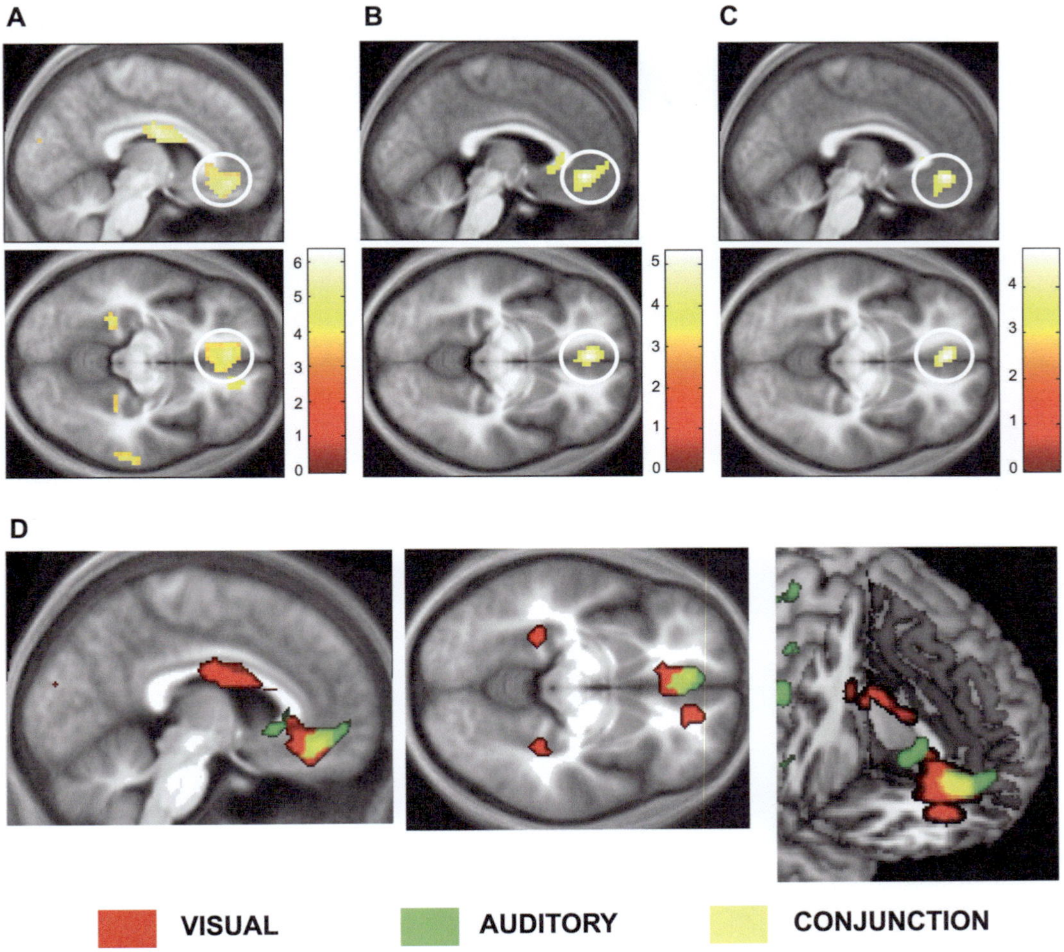

Fig. 3.28 Statistical parametric maps rendered onto averaged anatomical sections... (A) Visually Beautiful > Visually Ugly, (B) Musically Beautiful > Musically Ugly and (C) the results of a conjunction analysis for Visually Beautiful > Visually Ugly and Musically Beautiful > and Musically Ugly. Upper row shows activity in mid-saggital sections and the middle row in horizontal sections of the brain. (D) shows the overlap in zones within the medial orbito-frontal cortex (mOFC) activated by visually beautiful (red), musically beautiful (green) stimuli, and the overlap between the two activations (yellow). Tomohiro Ishizu and Semir Zeki. 2011. Cortical Activation correlating with the experience of beauty. Toward A Brain-Based Theory of Beauty. *PLoS ONE 6(7): e21852*. https://doi.org/10.1371/journal.pone.0021852. Reviewed February 6, 2023. Open access article

Also with an experimental approach, the Max Planck Institute for Empirical Aesthetics research on the cognitive and affective mechanisms involved in aesthetic evaluation, perception, and experience. These include neural, physiological and behavioural correlates in art forms as different as painting, music, literature, film, architecture and design, among others. Interestingly, the Institute has a multi-purpose event space, the ArtLab, designed to facilitate research on the individual experiences of an audience attending live presentations: synchronous registration of acoustic and optical data, facial expression, peripheral physiological and brain measures (Max Planck Institute for Empirical Aesthetics n.d.).

In a formal mathematical attempt to evaluate the aesthetic value of a piece of art, Alexander Huston and Joseph Huston (2014) gather the features or attributes (*qualia*) involved in the aesthetic experience as a weighted sum (Fig. 3.30).

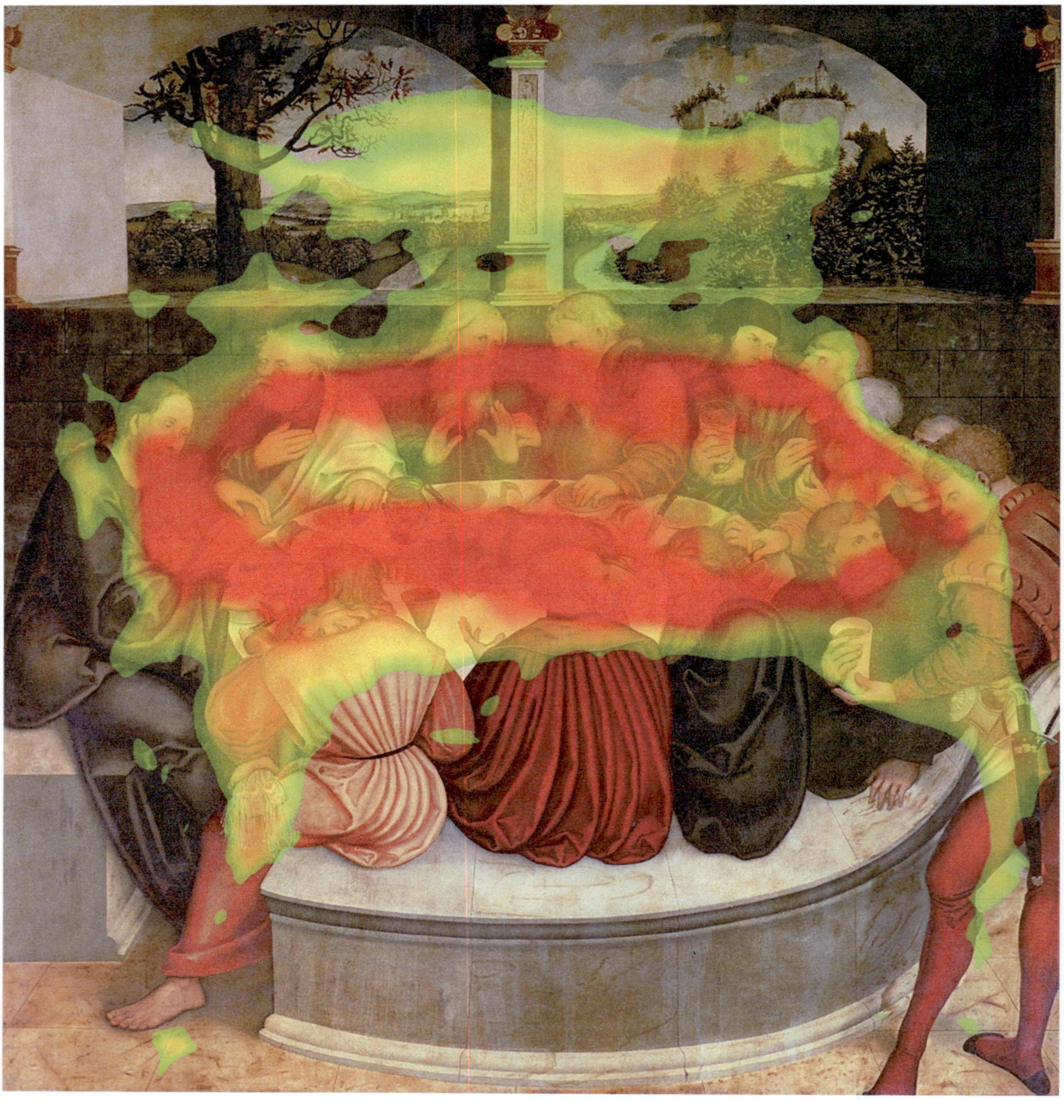

Fig. 3.29 Zoya Dare and Rosa Sancarlo. 2018. *Saccade Visualization over Cranach's Last Supper*. Image courtesy of CReA Lab, University of Vienna

$$E = \sum_{i=1}^{n} G_i w_i Q_i$$

Fig. 3.30 General equation for the subjective evaluation of an image where *E* (subjective overall quality) is a function of the number (*n*) of inherent *qualia* (*Q*) considered for the specific artwork. Each *quale* has a value (*g*), which is a measure of its *integrity* or *goodness*. A value weight (*w*) determines its relative *importance* for the work (Huston and Huston 2015, p. 481)

This formal analysis of an artwork includes basic *qualia* like colour or size, or more emotionally-loaded elements such as novelty, conflict, disgust or desire. Likewise, other elements such as perspective and balance, or poetic, ideological or spiritual characteristics can be considered. The authors apply this formalisation to different aesthetic domains, such as architecture, ancient, classical and modern art (Huston and Huston 2015).

Painter Robert Pepperel from the Cardiff School of Art & Design developed a remarkable technique based on visual indeterminacy, that is, when an image defies immediate or easy recognition. This attribute has been exploited by several artists throughout history to perplex audiences and constitutes a precedent of artistic movements such as Impressionism (Pepperell and Ishai 2015). Pepperel's painting technique consistently produces this indeterminacy effect (Fig. 3.31). His interest in the sensorial perception of art brought him to study the neural correlates of visual indeterminacy. Together with neuroscientist Alumit Ishai at the Max Planck Institute of Biological Cybernetics, they have found that amongst other behavioural differences, some of the brain areas activated by the view of indeterminate images differ from recognizable images, recruiting "higher cognitive functions, such as attention, visual imaginary and memory retrieval" (p. 152).

Nonetheless all these efforts, and regarding the subjective status of aesthetics in art, several questions persist. Why are certain works of art valued over others? What accounts for subjective differences? On the other hand, not all works of art are beautiful, and not everything beautiful constitutes a work of art (Nöe 2011). It seems that despite the efforts by scientists to answer these and similar questions, an understanding of subjective aesthetic experience continues to be elusive even for neuroaesthetics.[8]

In sum, this art-science category is driven by scientific purposes where the phenomenon studied belongs to the realm of art. The type, definition, or scope of the art phenomenon under study is determined by the scientific researcher, while it is the artist who may produce particular or unprecedented cases of artistic manifestations, as is the case of the indeterminate figures by Pepperel or Jung's drawings of the "unconscious". The conclusions of a scientific authority may be verified or substantiated through these works of art while the media platforms often used are those most commonly associated with science.

f. Works that encourage and foster creativity and innovation by exploring the relation between the arts, technology and science

Since the 1960s, efforts have been made to include artists in the development of innovative, new technologies. With the integration of cybernetics, kinetics and other advanced technologies in cutting-edge artworks, interest arose on the part of engineers and technicians to integrate the arts into the scientific revolution of the twentieth century.

In 1967 Billy Klüver, an electrical engineer from Bell Telephone Laboratories, together with artist Robert Rauschenberg and other artists and scientists, founded in New York a collaborative platform called *Experiments of Art and Technology, EAT*. Through exhibitions, performances and symposia, they expressed the urgency of creating a new consciousness and sense of responsibility associated with the crossover between art and technology. They foretell a new form of innovation where technology and art will precipitate a cultural revolution. In congruence with this aim they organised a series of performances, the *9 Evenings: Theatre and Engineering*, to facilitate communication and collaboration between engineers and artists (whom at the time had no professional nor social ties). This collaboration included local governments and interdisciplinary scientific research centres, the participation of professional artists, the media, in partnership with industry (Shanken 2005). Artists used the prototypes developed by the technology industry to deliver meaningful feedback as a thoughtful user, as opposed to passive consumers.

Other projects like *Arte de Sistemas* (1969) developed in Buenos Aires, or *Cybernetic Serendipity,* exhibited at the Institute of

[8] For a critical view on Neuroaesthetics see Ball (2013).

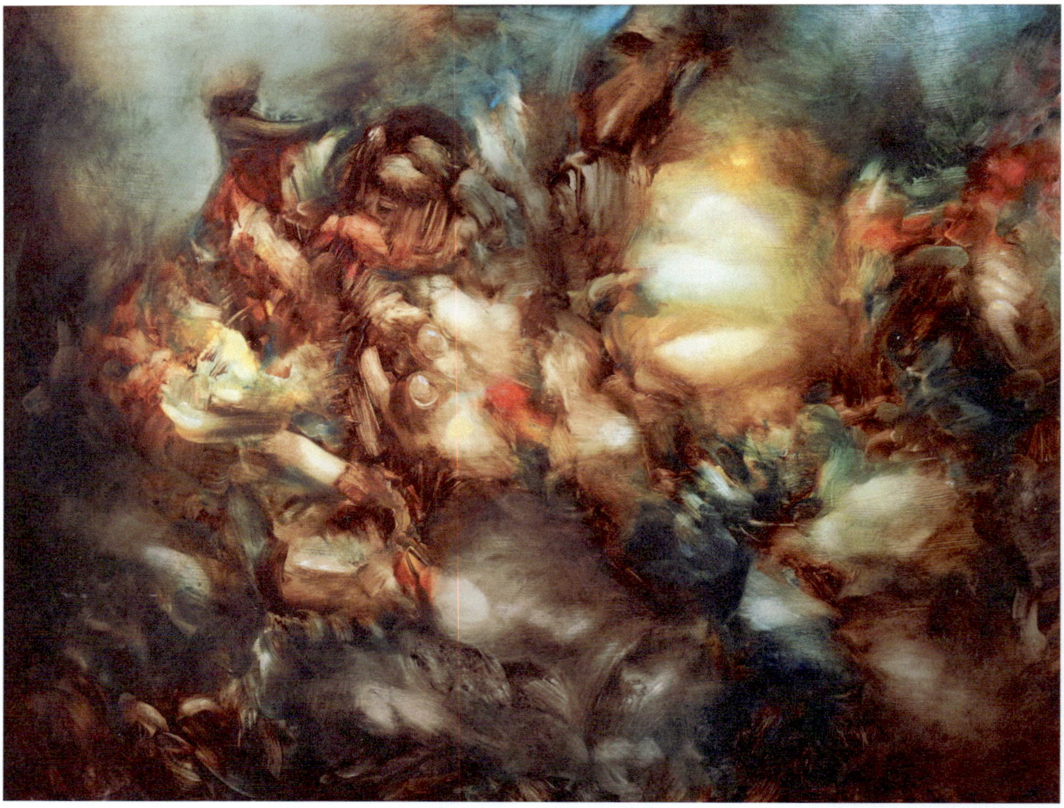

Fig. 3.31 Robert Pepperell. 2005. *Paradox 1.* Oil on Panel, 45 × 60 cm. Image courtesy of the artist

Contemporary Arts (1968) in London. Exploring the role of the machines used to create artwork, it raised crucial questions which influenced the construction of intelligent systems today. What happens to the observer when operating a machine and how can an observer's state modify the machine's internal structure?

In the 1990s, Dana Plautz, a documentary filmmaker, producer and professor of new media, worked in Intel's Multimedia Lab mediating between the engineering team and the creative department. One of her conclusions was that Intel should teach creative thinking in universities, giving rise to the Intel Art Entertainment Committee (IAEC), of which Plautz was chairman (Lee and Plautz 2006). In a similar line, the XEROX Parc was carried out in Palo Alto, California during the 1990s, engaging scientists and artists in a system of long-term residencies. Artistic pieces related to artificial life, telepresence, multimedia and immersive environments were developed. Xerox brought together artists with its in-house team of computer scientists, anthropologists, philosophers, linguists, interface designers, mathematicians and cryptographers and documented their findings. This space allowed artists to participate in the development of digital media.[9]

With a more public, social and cultural character, the ARS Electronica Festival and Exhibition Space (Fig. 3.32) began in 1979 in Linz, Austria and continues today as the world's largest art and technology exhibition. It is a hybrid model between business and the fields of culture and art. It has an annual competition system plus a series of other activities that include graduate programs in the use of information and communications technologies in different fields of art and design. It also has a museum and a residency program for educational purposes, and stimulates projects carried out by children.

[9] See Ref. Craig (1999).

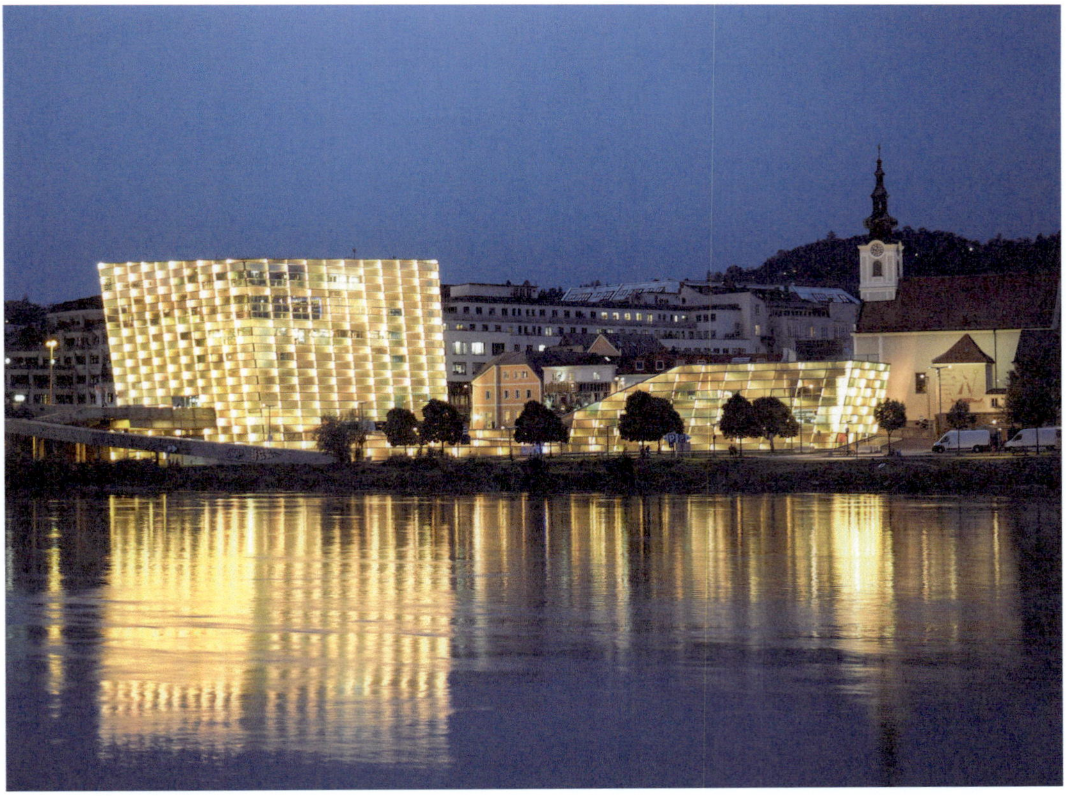

Fig. 3.32 Ars Electronica Centre. 2019. Photograph: Martin Hieslmair

The Massachusetts Institute of Technology (MIT) has a long tradition of connecting the worlds of art, science and technology inside its academic departments. In 1967 Bauhaus artist György Kepes created the Center for Advanced Visual Studies at MIT to bring together artists, scientists and engineers, pioneering the use of frontline technology in disruptive artistic projects. The MIT Media Lab, opened in 1985, is a research laboratory dedicated to projects in the convergence of technology, media, science, art, and design. According to their web portal (MIT Media Lab 2023) they have influenced various areas, such as emerging technologies for learning, entertainment, wearable computing, and sensor networks. It has been involved in the biological sciences and human–machine interface, innovating new materials and technologies for health and wellbeing. Today, it is focused on social, and global challenges in which ethics, inclusivity, sustainability, and the justice of technological development are key.[10]

At the end of the twentieth century, several European cities saw the implementation of different Media labs inspired by social and/or countercultural causes, with derivatives such as the *hack labs*, linked to activism and the DIY (Do It Yourself) movement (Tanaka 2011). In Germany, the Center for Art and Media Technology was created in Karlsruhe *ZKM* in 1997, where Hans Peter Weibel (2006) attempted to continue in the Bauhaus tradition. ZKM hired the mathematical physicist Hans H. Diebner who worked using an approach to knowledge that he called Performative Science (Diebner 2006). With a community focus, The Matadero Medialab is an institutional platform for research, creation and experimental production based in Madrid. Since

[10] See *A brief history of the Media Lab* www.media.mit.edu/about/history/, Reviewed on March 19th, 2023.

2002 it has addressed contemporary challenges linked to culture, the city and globalisation through public engagement and critical reflection, using digital tools and the talents of artists, designers and scientists, across different disciplines, institutions, organisations and industries.

As an examples of technologies developed from artistic purposes, *Ulises I* (Fig. 3.33), is a project from 2011 directed by Mexican artist Juan José Díaz Infante, together with the Mexican Space Collective, composed by Cera (rock band), Arturo Márquez (composer),

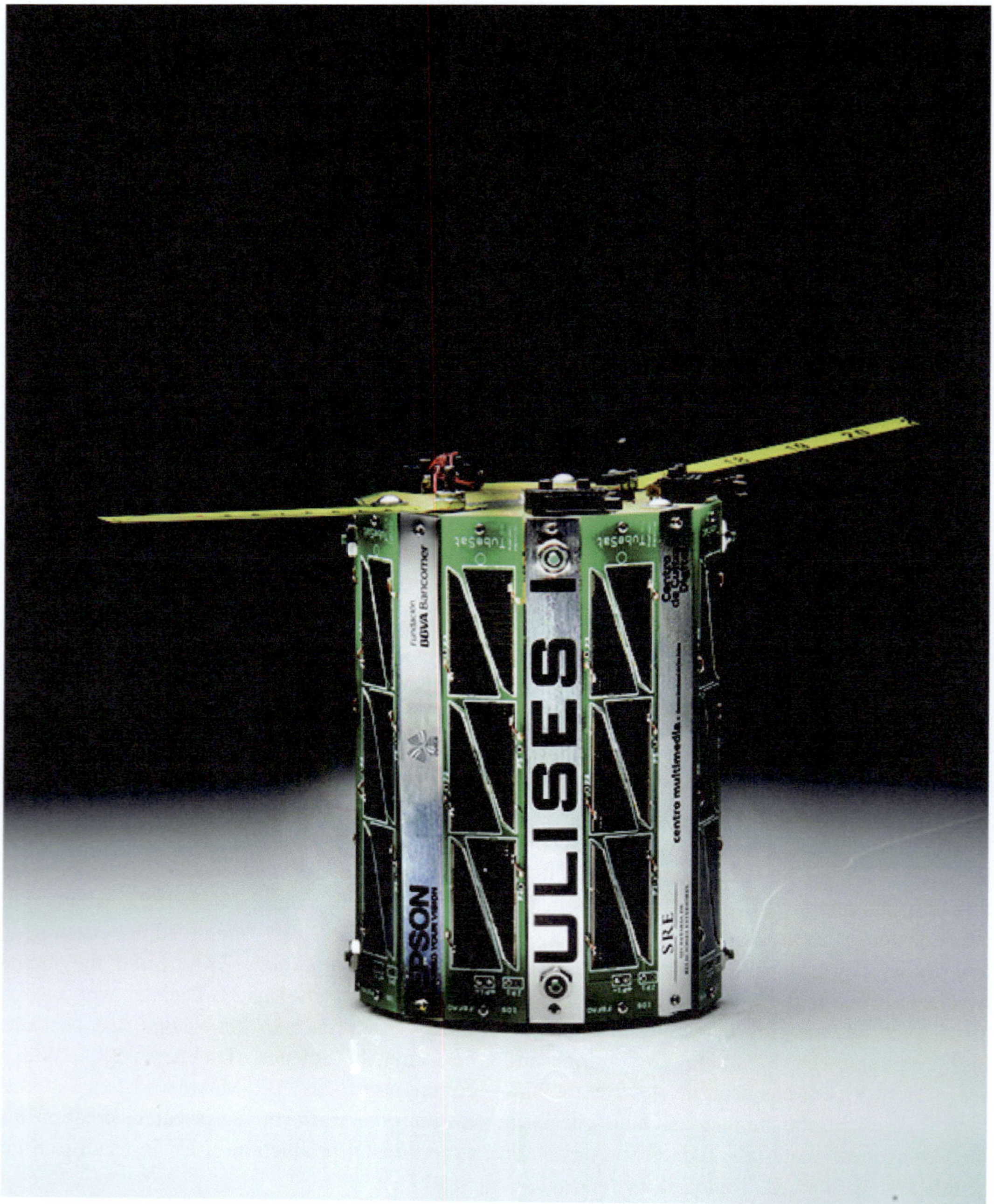

Fig. 3.33 Juan Jose Díaz Infante. 2017. *Ulises I*. Nano satellite. Image courtesy of the artist

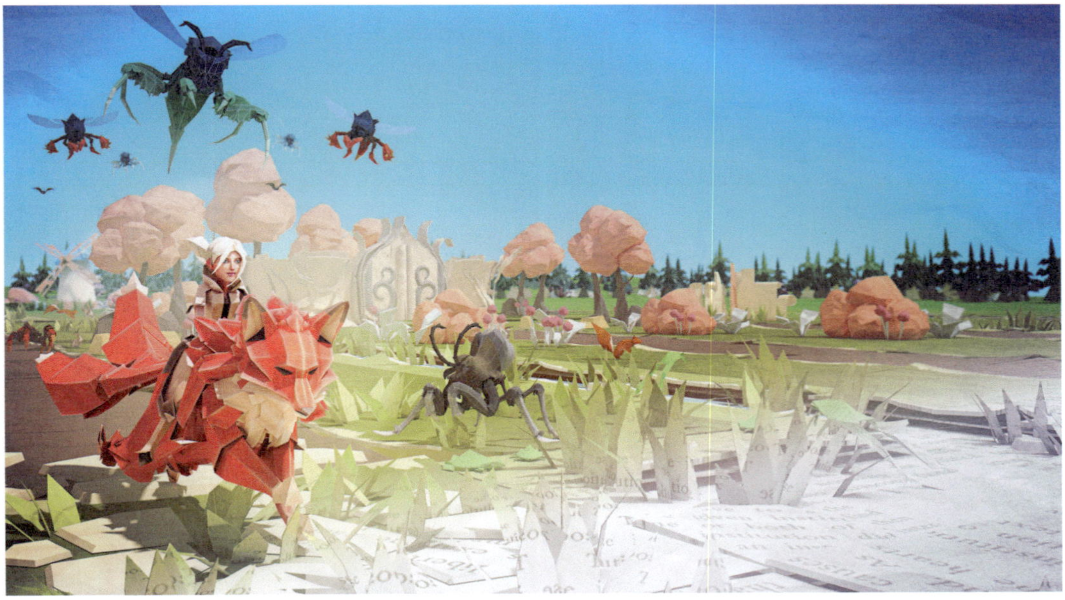

Fig. 3.34 Fishing Cactus, the developer. 2016. *Epistory plain*. GNU Free Documentation License, Version 1.2 or any later version published by the Free Software Foundation. https://commons.wikimedia.org/wiki/File:Epistory_plain.png. Reviewed May 9, 2023. Public domain

Hugo Solís (electronic artist), Omar Gasca (poet and composer), Juan Galindo and Julio Zaldívar (engineers in robotics, department of CENART's multimedia centre), Teresa Bordona (poet and doll artist) and Fernando Castro (photographer and philosopher). Diaz Infante defines *Ulises I* as an art project launched into space.[11] Its aim is to generate a poetic mission in orbit, launched from a country that does not even participate in the space race.

Another example in this category are video games, some of which are programmed and designed by artists, like the art collective JODI during the 1990s. Computer games have been a venue for the display of artistic approaches, with deconstruction, subversive gestures and the infusion of new meanings. Video games now have an immense participation in the entertainment industry[12] with one of three people on the planet playing them. They are complex hobbies in which several programming languages configure logical and imaginary worlds on a digital screen (Fig. 3.34). This industry is always testing new technologies and innovative designs to challenge the user. More recently, with the inclusion of artificial intelligence, digital games have become products that can stimulate and record different cognitive actions and reactions.

Finally, one of the most well-known, open-source, low cost digital electronic hardware systems, Arduino (Fig. 3.35), was created in 2005 by Massimo Banzi and David Cuartielles based on the work of Hernando Barragán. Arduino uses a reprogrammable microcontroller with a series of pins that allows easy connection with different devices as inputs and outputs (e.g. sensors and actuators). Artistic developments using this digital board have proven its flexibility and potential to several applications, driven by complex interactive art pieces and industrial prototypes.

As has been shown, this category is quite symmetrical as the ideas and practices of science, technology and art intermix, feed upon and feedback to each other, achieving diverse results that are functional for different purposes

[11] Email sent by Juan Jose Díaz Infante on May 10, 2023, to Ignacio Nieto.

[12] The earnings in 2021 were US$193 billion. Kellen (2021).

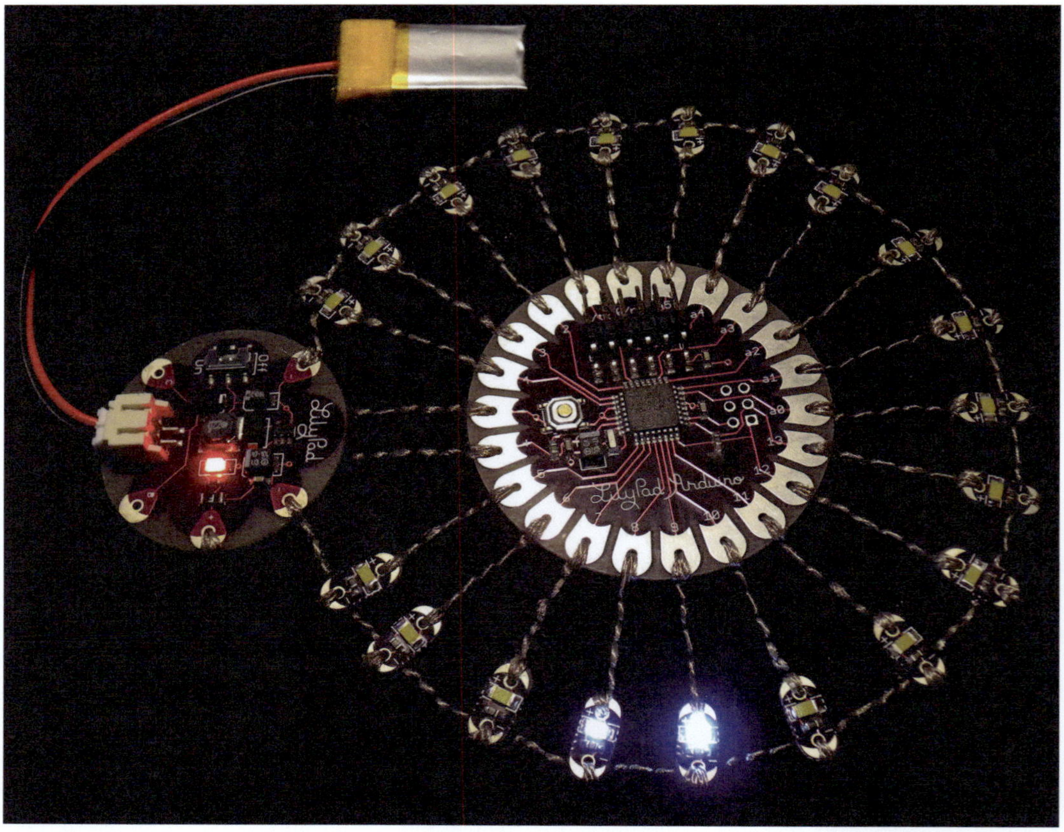

Fig. 3.35 Iwasaki Osamu (2009) *Lilypad Arduino version, connected to leds, powered by an external battery*. https://commons.wikimedia.org/wiki/File:Arduino_Lilypad_LED_circle.jpg Reviewed May, 2023. Public domain

and valued by a variety of communities or businesses. To the extent that the goal of the integrated work is clear and bounded by a very specific purpose, be it artistic, scientific, economic or community-based, it is likely that a more clearly symmetrical integration will occur. Its media platform is as diverse as innovative products or services may be.

In the next chapter, a set of contemporary art works will be presented and described in some detail by their own creators. Almost all of them were developed in what have been identified as a zone of "symmetrical integration", where works that display a prevailing art purpose are set in a scientific framework, or exactly the reverse. In both cases, as this chapter intimates, this integration of contemporary art and cutting-edge science can produce significant results for its creators and for audiences. These works enrich current knowledge and have the potential to transform our vision of the world and thus our actions.

References

Alghamdi MA, Ziermann J M and Diogo R (2016) An untold story: The important contributions of Muslim scholars for the understanding of human anatomy. *The Anatomical Record*. 300:986–1008, pp. 979–1170 VVC2016 WILEY PERIODICALS, INC. https://anatomypubs.onlinelibrary.wiley.com/doi/epdf/10.1002/ar.23523. Reviewed 14 February 2023.

Ball P (2013) Neuroaesthetics is killing your soul. *Nature*, https://doi.org/10.1038/nature.2013.12640.

Catts O and Zurr I (2014) Growing for different ends. *The International Journal of Biochemistry & Cell Biology*. Volume 56, Pages 20–29.

Craig H (1999) *Art and Innovation*. MIT Press, Cambridge.

Becker H (1982) *Art Worlds*. Berkeley and Los Angeles: University of California Press, cited in Kimmel, M.S.

(1983). [Review of Art Worlds., by H. S. Becker]. *American Journal of Sociology*, 89(3), 733–735. http://www.jstor.org/stable/2779026

Browning Kellen (2021) "Game world celebrates at its Oscars," *The New York Times*, 11 December 2021, via Factiva.

Yetisen AK, Davis J, Coskun AF, Church GM and Yun SH (2015) Bioart. *Trends Biotechnol.* Dec;33(12):724–734. doi: https://doi.org/10.1016/j.tibtech.2015.09.011. PMID: 26617334.

Bibliography

Badagliacca V (2021) Life inside a system: eco-critical stances in the art of Luis Fernando Benedit, 1968–72. *Burlington Contemporary Journal*. November https://contemporary.burlington.org.uk/journal/journal/life-inside-a-system-eco-critical-stances-in-the-art-of-luis-fernando-benedit-196872. Reviewed March 6, 2023.

Borel A (1983) "Mathematics: Art and Science". In *The Mathematical Intelligencer*. Vol. 5, No. 49 Springer-Verlag, New York.

Catts O, Zurr I & Ben-Ary G (2001) *Semi-living Worry Dolls* https://tcaproject.net/portfolio/worry-dolls/. Reviewed April 2023.

Daston L and Galison P (2010) *Objectivity*. Zone Books.

Davis J (1996) Microvenus. *Art Journal*. Vol. 55, No. 1 Contemporary Art and the Genetic Code. pp. 70–74.

Ede S (2005) *Art & Science*. I.B. Tauris.

Eigler D and Schweizer E (1990) Positioning single atoms with a scanning tunnelling microscope. *Nature* 344, 524–526. https://doi.org/10.1038/344524a0.

Fechner G (2013) *Vorschule der Aesthetik* (Cambridge Library Collection - Art and Architecture). Cambridge: Cambridge University Press. https://doi.org/10.1017/CBO9781139854580.

Feyerabend P K (1994) Art as a Product of Nature as a Work of Art. In *World Futures: The Journal of New Paradigm Research,* 40:1–3, 87–100. https://doi.org/10.1080/02604027.1994.9972421.

Foucault M (1966) *Las palabras y las cosas. Una arqueología de las ciencias humanas.* Buenos Aires, Siglo Ventiuno Editores.

Freedberg D and Gallese V (2007) Motion, emotion and empathy in esthetic experience. *Trends in cognitive sciences*, 11(5), 197–203.

Garcia T (2014) *Form and Object. A Treatise on Things.* Edinburgh University Press.

Gell A (2016) *Arte y Agencia. Una Teoría Antropológica.* SB Editores, Buenos Aires.

Grillmayr J (2020) "The Many-Layered Cake of Science Fiction". *In Arts in Context – Kunst, Forschung, Gesellschaft.* Gartmann T & Pauli C (Hg.). E-Book (PDF) Open Access. https://www.transcript-verlag.de/978-3-8376-5322-9.

Guagnini N (2013) "Retroalimentación en el Amazonas". In Gonzáles J and Ramirez Moyao A. *Juan Downey. Una Utopía de la Comunicación*. Museo Tamayo Ciudad de México.

Huston A J and Huston J P (2015) Aesthetic Evaluation of Art: A Formal Approach. In Huston JP, Nadal M, Mora F, Agnati LF and Cela-Conde CJ (eds.). *Art, aesthetics and the brain.* Oxford University Press. https://doi.org/10.1093/acprof:oso/9780199670000.001.0001.

International Human Genome Sequencing Consortium (2001) Initial sequencing and analysis of the human genome. *Nature* 409 (6822): 860–921. https://doi.org/10.1038/35057062.

Ishizu T and Zeki S (2011) Toward A Brain-Based Theory of Beauty. *PLoS ONE* 6(7): e21852, https://doi.org/10.1371/journal.pone.0021852.

Ivanova M (2017) Aesthetic values in science. *Philosophy Compass*. Volume12, Issue10. https://doi.org/10.1111/phc3.12433.

Jacobsen T (2006) Bridging the Arts and Sciences: A Framework for the Psychology of Aesthetics. *Leonardo*, Vol. 39, No. 2, pp. 155–162.

Kac E (2007) "Life Transformation - Art mutation". In Kac E (ed.) (2007) *Signs of life*. Bio Art and beyond. MIT Press.

Kandel E (2016) R*eductionism in Art and Brain Science: Bridging the Two Cultures*. Columbia University Press.

Kemp M (1998) Kendrew constructs; Geis gazes. *Nature* 396, 525. https://doi.org/10.1038/25019.

Kemp M (2003) Science in culture. *Nature* 423, 483. https://doi.org/10.1038/423483a.

Kemp M (2005) From Science in Art to the Art of Science. *Nature* 434, 308. https://doi.org/10.1038/434308a.

Latour B (1987) Science in action: How to follow scientists and engineers through society. Cambridge, MA: Harvard University.

Lavin I (1994) "The Art of Art history". In *World Futures: The Journal of New Paradigm Research*, 40:1–3, 13–25. https://doi.org/10.1080/02604027.1994.9972413.

Lee N and Plautz D (2006) Interview with Dana Plautz. *Comput. Entertain.* 4, 1 (January), 2–es. https://doi.org/10.1145/1111293.1111297.

Max Planck Institute for Empirical Aesthetics (n.d). *What Is the ArtLab?* https://www.aesthetics.mpg.de/en/artlab/information.html.

Macneill P U and Ferran B (2011) Art and bioethics: Shifts in understanding across genres. *Bioethical Inquiry*, 8, 71–85.

Martinoya C and Joël N (1968).The Chromatic Abstractoscope: An Application of Polarized Light. *Leonardo,* 1 (N°2), 171–173.

Maturana H, Mpodozis J (1987) Perception: behavioral configuration of the object. *Arch. Biol. Med. Exp;* 20(3/4): 319–24.

McMullen K (2005) Experimental physics, experimental art. Artists on Science. *Nature* 434, 308. https://doi.org/10.1038/434310a.

Mit Media Lab (2023) *About the Lab History* https://www.media.mit.edu.

Mitchell R (2010) *Bioart and the Vitality of Media.* University of Washington Press.

Morris P J T (2015) The Matter Factory: A History of the Chemistry Laboratory. London: Reaktion Books.

Noah Y (2017) *The Mozart in the machine: Computers may become better artists than humans.* Stuff. May 08, 2017. https://www.stuff.co.nz/technology/digital-living/92322848/the-mozart-in-the-machine-computers-may-become-better-artists-than-humans.

Noë A (2011) *Art and the Limits of Neuroscience. The Opinion Page.* The New York Times Company. https://archive.nytimes.com/opinionator.blogs.nytimes.com/2011/12/04/art-and-the-limits-of-neuroscience/. Reviewed on February 23, 2023.

Norton B M (2020) Aesthetics, Science, and the Theater of the World. *New Literary History*, Volume 51, Number 3, pp. 639–659.

Pepperell R and Ishai A (2015) "Indeterminate artworks and the human brain". In Huston JP, Nadal M, Mora F, Agnati LF and Cela-Conde CJ (eds.). *Art, Aesthetics, and the Brain.* Oxford, University Press.

Prodger P (1998) "Illustration as Strategy in Charles Darwin's The Expression of Emotions in Man and Animals". In Lenoir T (ed.) (1998) *Inscribing Science: Scientific Texts and the Materiality of Communication.* Stanford, CA: Stanford University Press.

Ramachandran V S and Hirstein W (1999). The Science of Art. A Neurological Theory of Aesthetic Experience. *Journal of Consciousness Studies*, 6, No. 6–7, pp. 15–51.

Rosenberg R and Klein C (2015) "The moving eye of the beholder: Eye tracking and the perception of paintings". In Huston JP, Nadal M, Mora F, Agnati LF and Cela-Conde CJ (eds.) (2015). *Art, aesthetics and the brain.* Oxford University Press. https://doi.org/10.1093/acprof:oso/9780199670000.001.0001.

Sancarlo R, Dare Z, Arato J and Rosenberg R (2020) Does Pictorial Composition Guide the Eye? Investigating Four Centuries of Last Supper Pictures. *Journal of Eye Movement Research* 13. DOI: https://doi.org/10.16910/jemr.13.2.7.

Scott J (2006) *Artists in Lab. Process of Inquiry.* Springer; Auflage.

Searle J R (1979) *Expression and Meaning. Studies in the theory of speech acts.* Cambridge University Press. https://doi.org/10.1017/CBO9780511609213.

Searle J R and Vanderveken D (1985) "Speech Acts and Illocutionary Logic". In: Vanderveken D (ed.) (1985) *Logic, Thought and Action. Logic, Epistemology, and the Unity of Science*, vol 2. Springer, Dordrecht. https://doi.org/10.1007/1-4020-3167-X_5.

Shanken E (2005) Artists in Industry and the Academy: Collaborative Research Interdisciplinarity Scholarship, and the Interpretation of Hybrid Forms. *Leonardo* Volumen 38 N° 5, pp. 415–418.

Shankland S (2009) *IBM's 35 atoms and the rise of nanotech.* CNET. September 18 https://www.cnet.com/culture/ibms-35-atoms-and-the-rise-of-nanotech/.

Stettler R (2006) "Perception – Translation Transformation. Artists in Lab". In Scott J (ed.) *Process of Inquiry.* Springer, Auflage.

Tanaka A (2011) Situating within Society. Blueprints and Strategies for the Media Labs. *Blueprints for the Future for the Lab.* Eindhoven: Baltan Lab Publishing.

Tooley R V (1979) *Tooley's Dictionary of mapmakers.* New York: A. R. Liss.

Udías A (2020) Athanasius Kircher and Terrestrial Magnetism: The Magnetic Map. *Journal of Jesuit Studies.* Volume 7: Issue 2.

Villoro L (1993) The idea of nature in the Renaissance. *Science*, no. 29, January-March, pp. 74–80. https://www.revistacienciasunam.com/pt/178-revistas/revista-ciencias-29/1666-la-idea-de-la-naturaleza-en-el-renacimiento.html. Reviewed January 29, 2023.

Weibel P (2006) Preface: "A Museum Mission". In Diebner H H (2006) *Performative Science and Beyond. Involving the Process in Research.* Springer Wien/New York ZKM Center for Art and Media Karlsruhe.

Wildevuur S (2009) *Invisible Vision. Could Science learn from the Arts?* Nederlands Tijd-schrift voor geneeskunde. Bohn Stafleu van Loghum.

Wiles K (2017) The Map: The Principal Mountains and Rivers of the World, 1829. *History Today* Volume 67 Issue 5.

Zeki S, Romaya J P, Benincasa D M T and Atiyah M F (2014) The experience of mathematical beauty and its neural correlates. *Front.Hum.Neurosci.* 8:68. https://doi.org/10.3389/fnhum.2014.00068.

4 Selected Art-Science Works

Abstract

This chapter presents 22 works by art-scientists who describe them providing valuable information and insight into their material and expositive characteristics, and their underlying theoretical bases. Each work is shown in full colour images. This selection of works was chosen to expose the symmetrical authority of these art-science integrations, and the inspirational value and knowledge generating potential they demonstrate.

Keywords

Art-scientists · interdisciplinarity · Science communication · Bioart

4.1 Introduction

In this chapter, 22 art-science works are presented, chosen from active art-scientists from various countries who were invited to participate in this book and kindly accepted. The selection of the set entailed a process akin to a curatorship. Each work is described by their creators in a free exercise, rather uncommon in the world of art, and is accompanied by images the authors provided. Some are professional artists trained in science, many of whom hold postgraduate degrees, while others are professional scientists who have ventured into the world of contemporary art, supporting projects or directly leading contemporary artworks that use and depend on their research experience and capabilities.

One of the keys to the selection of these art-science works is the balanced role of both science and art they display in terms of the process of creation and the work itself, in a kind of "interinfluence". Each piece embodies a symbiosis of an inter and transdisciplinary nature challenging the process of knowledge generation. These works address a wide range of issues: global crises, education, the brain, the body, molecules and cells, the self and otherness, the living and nonliving, the human and non-human, the future, nature, empathy, compassion with the neglected of the world, and the excitement of exploration and discovery.

Readers are invited to explore these art-science integrations, particularly in terms of the different ways the works articulate art and science and the symmetrical/asymmetrical art-or-science authority relations they express as a way to include new sensibilities in the research practice. The 22 works are enumerated in Table 4.1.

© The Author(s), under exclusive license to Springer Nature Switzerland AG 2024
M. Velasco and I. Nieto, *The Art-Science Symbiosis*, https://doi.org/10.1007/978-3-031-47404-0_4

4.2 Art-Scientist, Selected Works

Table 4.1 This list provides information about the works presented in this Chapter. The order of appearance is alphabetically arranged according to the surname of one of the authors of each work

	Selected work title/format	Artist(s)/scientist(s) Nationalities	Year of showing
1	*Immersion of the Henri Poincaré Institute in the R^3/Z^3 space* Sound installation	Pierre Berger, Vincent Martial (France) and Jimena Royo-Letelier (Chile)	2017
2	*Evolvable Yeast Automaton* Installation	Jean-Baptiste Boulé, Nicolas Desprat and Philippe Parreno France	2016
3	*That Which Lives in Me* Installation	Dmitry Bulatov and Alexey Chebykin Russian Federation	2011
4	*Formata* Hybrid Installation	PЯOTO-ALIEИ PЯOJECT (Head of the project: Juan M. Castro) Colombia–Japan	2020–2022
5	*2021 Kolam* Public space intervention	Shanthi Chandrasekar India–USA	2021
6	*Harmonic Dissonance* Interdisciplinary and interactive performance art installations	Matthias Oostrik, Suzan Tunca, Than van Nispen (The Netherlands) and Suzan Tunca (Turkey–Germany)	2018–2023 (ongoing)
7	*Nomadic Plant* Installation	Gilberto Esparza Mexico	2008–2013
8	*Integrative Illustration of Cells and Viruses* Illustrations	David Goodsell USA	2020
9	*Blood Wars* Installation/Performance	Kathy High USA	2010–2011
10	*Impulse* Installation	Yunchul Kim South Korea	2018
11	*Microbiome Rebirth Incubator* Installation	François-Joseph Lapointe and Marianne Cloutier Canada	2016–2023
12	*Plastic is Land* Digital photographs	Max Liboiron Canada	2018
13	*Primate Cinema: Apes as a Family* Film	Rachel Mayeri USA	2011
14	*Creatures* Graffitis	Andro Montoya Riveros Chile	2008–2021
15	*Solar Displacement* Performance	Špela Petrič Slovenia	2014
16	*Green, Grey or Dull Silver* Fieldwork intervention	Perdita Phillips Australia	2007–2008
17	*Inhabiting a Map* Computational graph	Michail Rybakov Russian Federation	2018
18	*Bee's* Installation	Susana Soares Portugal	2007–2009
19	*Crocheted hyperbolic planes* Crochet Installation	Daina Taimina Latvia–USA	2009–2014
20	*Living Architectures* Projection mapping	Manuel Théry France	2008, 2013
21	*Labor* Installation	Paul Vanouse USA	2019–
22	*Neuroportraits* Graphics	Nicholas Wade Scotland	2018

4.3 Art-Science Work Descriptions by Their Authors

In what follows, the authors describe and present images of their work, all of which were produced between 2007 and 2021. Readers will note a diversity of styles as the authors have expressed their own views.

4.3.1 *Immersion of the Henri Poincaré Institute in the R^3/Z^3 Space.* Pierre Berger, Vincent Martial and Jimena Royo-Letelier, 2017

By Jimena Royo-Letelier

A sound immersion in a mathematical space

This sound and visual work intervened in-situ the Henri Poincaré Institute in Paris (Fig. 4.1),

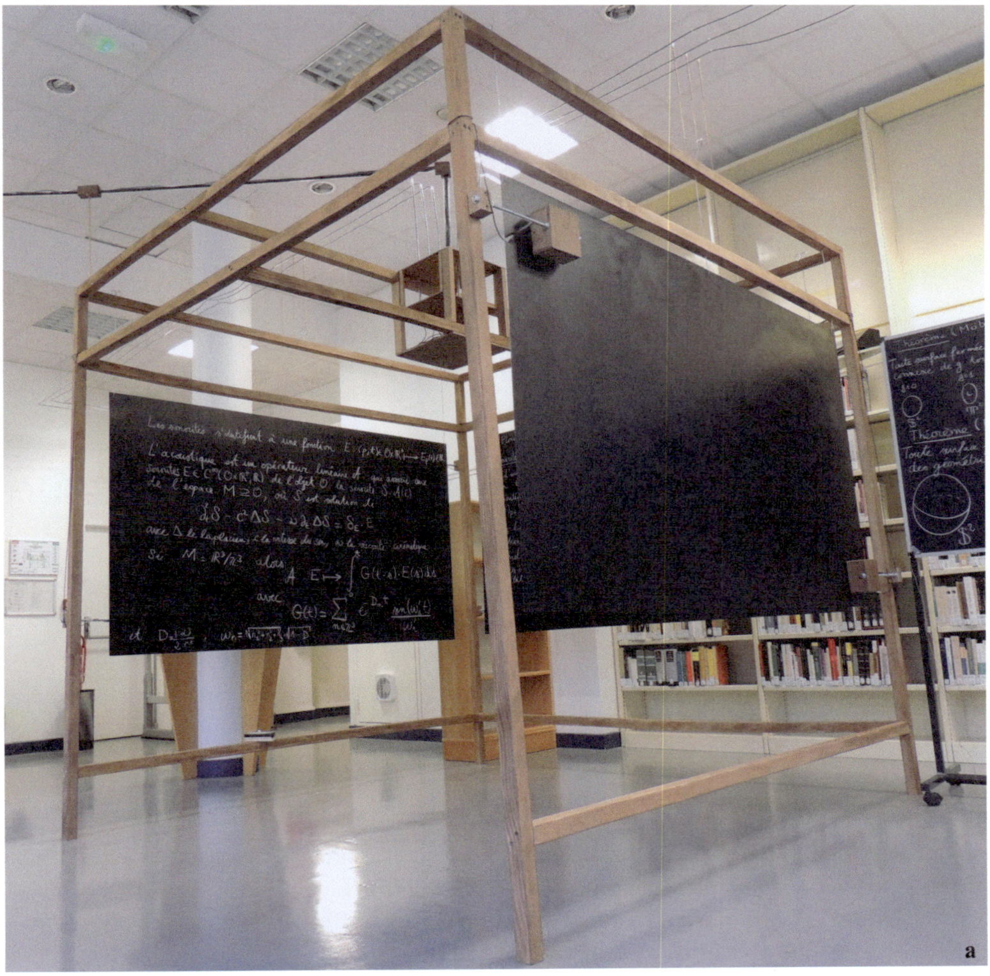

Fig. 4.1 Pierre Berger, Vincent Martial and Jimena Royo-Letelier. 2017. *Immersion of the Henri Poincaré Institute in the R^3/Z^3 space*. Henri Poincaré Institut, Paris, France. **a** General view. Picture credits Pierre Berger. **b** Contact microphones picking up building vibrations. Picture credits Pierre Berger. **c** Explanations of the device on the blackboard diffusing the calculated sounds. Picture credits Pierre Berger

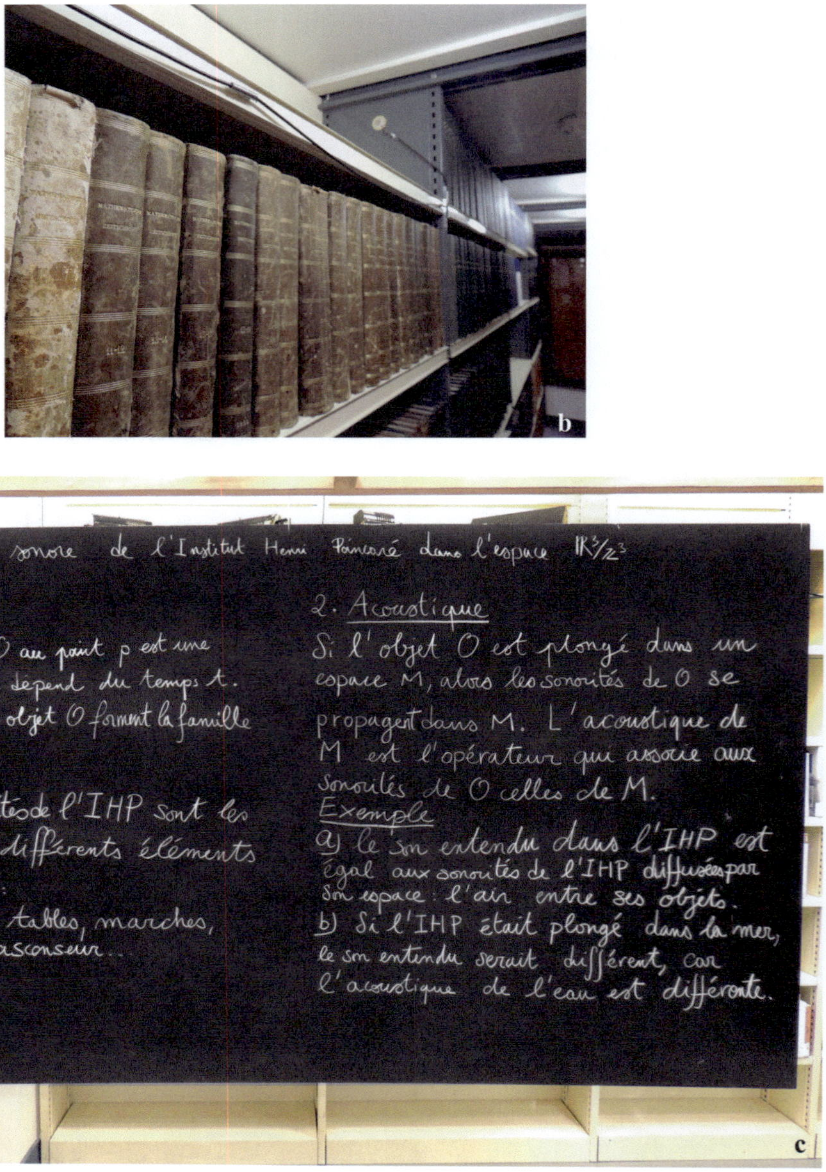

Fig. 4.1 (continued)

creating a sound immersion of its main building in an imaginary mathematical space. This space, called R^3/Z^3, does not exist in our reality, but the public is invited to explore it thanks to an electromagnetic, mechanical and computerised device that recreates the sound transmission that occurs there. The sounds resonate differently, the acoustics are different.

Dozens of contact microphones were installed throughout the building, capturing the different sounds that pass through it: people discussing ideas, writing on the blackboards, drinking coffee, taking books off the shelves, the noise of the ventilation ducts or the elevator going up and down. These were then transmitted to a computer device that recreated in real time the acoustics of the R^3/Z^3 space, to be finally restored sonically to the audience, through electromechanical transducers attached to suspended blackboards, which act as amplifiers-speakers, allowing viewers to be transported sonically to the mathematical space.

The sounds captured and returned in this way in the acoustics of the space allowed the public to perceive the sensations that researchers experience when they explore mathematical objects in their daily work. A work of forging intuition on abstract constructs, thanks to empirical exploration via mathematical calculations and computer simulations.

The algorithm implemented in the device of the work, which allows the calculation of the acoustics of the imaginary space, was developed by the authors after research that allowed them to understand how sound propagates in space. The work aims to share the sensations that emerged during the research process, the sensory aesthetic that the researchers faced and sought in the process of creation and mathematical research.

The immersion work was presented as part of the exhibition *Esthétopies*. This audio-visual exhibition, designed between mathematicians, computer scientists and artists, tries to share with the public the taste of the emotions experienced when facing a mathematical problem: first the intuition, then the discovery, and finally the mathematical formalisation, which allows the idea to be expressed clearly as a mathematical truth (a theorem with its proof, or new concepts). Art allows us to express the unspeakable part of the process: between blind intuition and the moment in which one apprehends and sees a phenomenon, without being able to express it. It is a fascinating and frustrating contact with a mathematical reality.

4.3.2 *Evolvable Yeast Automaton.* Jean-Baptiste Boulé, Nicolas Desprat and Philippe Parreno, 2016

By Jean-Baptiste Boulé and Nicolas Desprat

The project we carried out in collaboration with Philippe Parreno and his Atelier, includes four exhibitions, one occurring in a gallery (Gladstone Gallery, NY) and three in museum institutions (Tate Modern, London; Jumex Foundation, Mexico City; and Gröpius Bau in Berlin.

The goal of our collaboration with Philippe Parreno was to design an organic control for his installations, with the idea that the control should have essential features such as evolvability and memory in order to generate a continuously drifting loop of events choreographed by the artist. The basis of the project was to use a population of living microorganisms (a yeast culture) as a control centre (a brain) to organise an ensemble of objects and media (films, sounds and mobile elements), the whole composing an organism interconnected with an environment (the exhibit itself).

To this end we designed a connected bioreactor able to interact with an "environment". The exhibition's environment was composed of automatons, sounds and movies, organised in such a way that the bioreactor could receive numerous inputs that modified the environmental conditions within it through actuators affecting oxygenation, pH, temperature, the nature of the carbon source (the food) and the rate at which medium is renewed. In return, according to its physiological state the bioreactor sent feedback to the environment of the exhibit through algorithms that we specifically established for each exhibit. This feedback directly affected the dynamics of the exhibits, through interaction with automatons, lights and sounds (Fig. 4.2c).

Each exhibit relied on a different environment and narrative proposed by the artist, so each time we established a set of rules for the coevolution of the yeast culture and the exhibit to define how they could influence each other and evolve together during the time of the exhibit. Although each exhibit explored diverse narratives, the common aspect was that each one was composed of several "quasi objects" (named as such by the artist) whose dynamics he created initially. For some of these objects, the bioreactor could send data resulting from a transformation of the recording of various probes (RedOx state, acidification by cells, respiration levels, growth) and receive data from the exhibit in return, which would determine the state of the bioreactor. Each state could be defined as a set of growth conditions (media composition, speed of medium renewal, air influx, temperature, etc....).

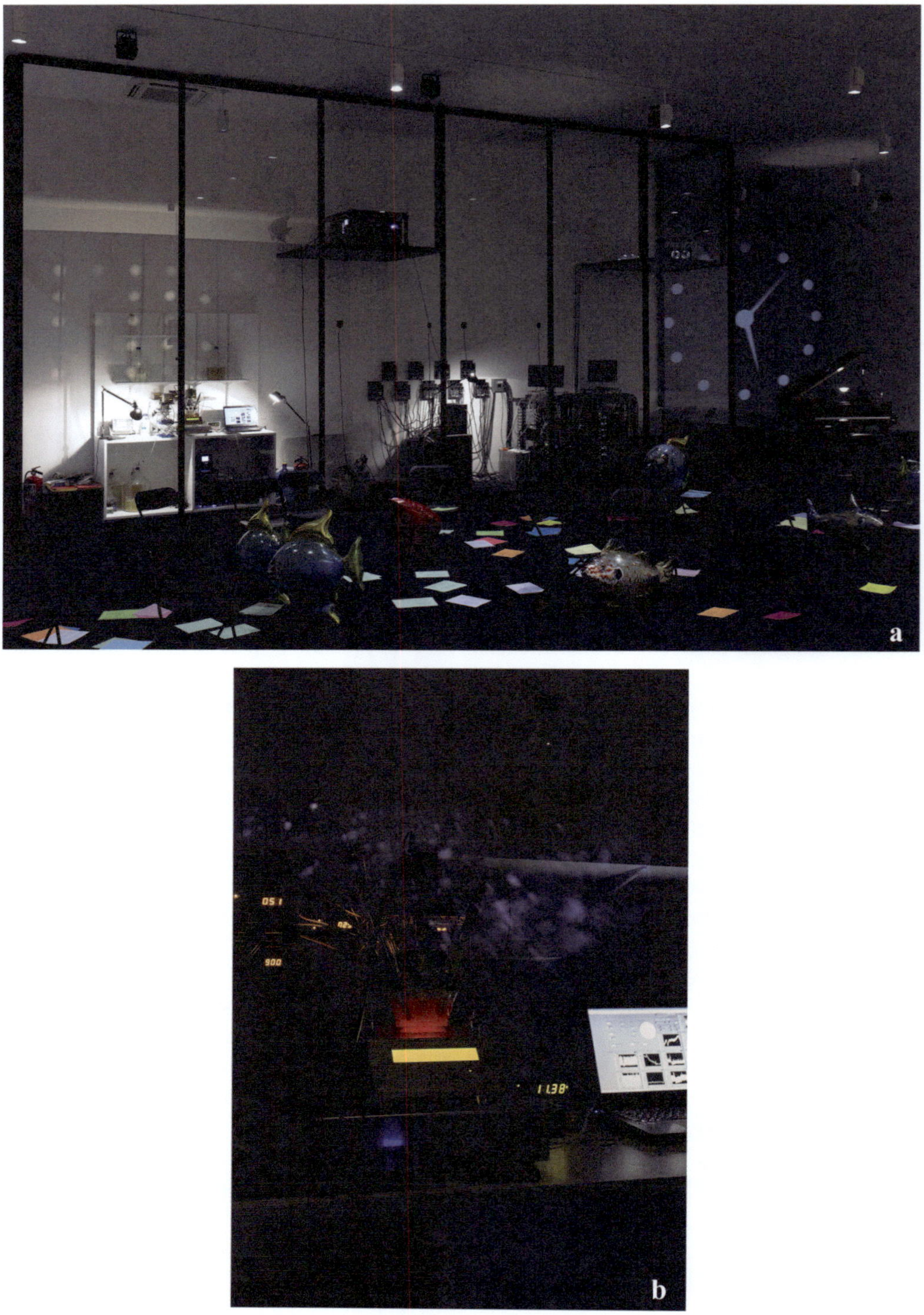

Fig. 4.2 Philippe Parreno and Gropius Bau. 2018. *Evolvable Yeast Automaton*. **a b** Exhibition view: Berlin. Courtesy of the artist, Pilar Corrias, London, Gladstone Gallery, New York/Brussels, and Esther Schipper, Berlin) Photo © Andrea Rossetti. **c** Philippe Parreno. 2017–18. *Evolvable Yeast Automaton. La Levadura y el Anfitrión*. Exhibition view. Museo Jumex, Mexico City. Courtesy of the artist, Pilar Corrias, London, Gladstone Gallery New York/ Brussels, and Esther Schipper, Berlin. Photo © Andrea Rossetti. **d** Jean-Baptiste Boulé. 2022. Schematics of the principle of the bioreactor-exhibit coupled automaton (pO2: partial oxygen pressure)

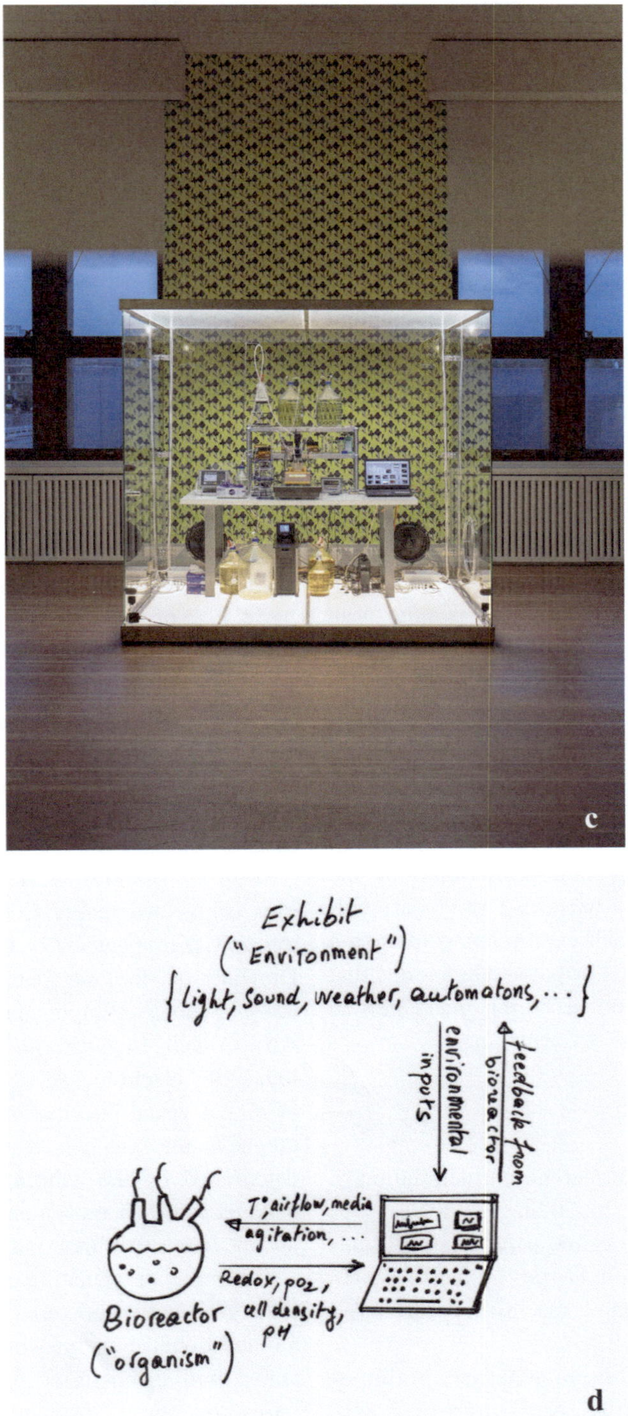

Fig. 4.2 (continued)

From a biological point of view, the difficulty of building such a bioreactor is the need to create several states that are sufficiently varied to allow a clear distinction between them (such that the output of the exhibit is clear), but also compatible so that the culture remains alive and healthy throughout the exhibits which span several months. In other words, we created compatible environments within the bioreactor while allowing the exhibit to create a narrative with a variable environment made of many states (or you could call them seasons, or biotopes) over a long period of time (one to six months) with minimal maintenance operation except replenishing media bottles regularly.

In this way, the exhibit consisted of a conversation between the bioreactor and the exhibit, or between an "organism" and its environment. At the scientific level, the exhibit is akin to an experiment investigating adaptation to random or predictable environmental variations.

By design, the experiment is not easily reproducible, so it does not meet the criteria inherent to scientific methodology. However, we recovered every input coming from the exhibit so as to allow the possibility to play back an entire exhibit during another iteration of the experiment. We also preserved the yeast culture resulting from each exhibit, so as to keep a genetic memory of the possible adaptations that occurred over the course of the yeast growth cycle in this unnatural environment.

References

- https://www.gladstone64.com/installation/6. Reviewed January 26, 2023.
- https://www.tate.org.uk/art/artists/philippe-parreno-9107/philippe-parreno-half-mechanic-half-organic-machine. Reviewed January 26, 2023.
- https://www.tate.org.uk/art/artists/philippe-parreno-9107/philippe-parreno-half-mechanic-half-organic-machine. Reviewed January 26, 2023.
- https://www.fundacionjumex.org/en/exposiciones/90-philippe-parreno-the-yeasts-and-the-host. Reviewed January 26, 2023.
- https://www.berlinerfestspiele.de/en/berliner-festspiele/programm/bfs-gesamtprogramm/programmdetail_237340.html. Reviewed January 26, 2023.

4.3.3 *That Which Lives in Me*. Dmitry Bulatov and Alexey Chebykin, 2011

By Dmitry Bulatov

This project (Fig. 4.3) investigates the transformative capacity of temporal rhythms offered by the non-living physical environment. The basic law of inert matter consists of reacting immediately to being affected, the reaction taking up the same rhythm and continuing through the same duration. With regard to living beings, it would be fair to say that the independence of their effect on surrounding matter is strengthened to the extent that they break free from the rhythms immanent to non-living matter.

That Which Lives In Me is an installation that uses augmented reality technologies to enable a dynamic rearrangement of real and virtual spaces. Drawing on the metaphor of a snail's "shell house", we have "augmented" real shells of giant African snails *(Achatina fulica)* with a certain analogue of "electronic aura"—an interactive layer of digital visual information. This information is output to the screen in real time, its appearance determined by the behaviour of the snails: the dynamics of their movements inside the terrarium and the intensity of their inter-communication.

To create an alienating effect, we made use of the snails' active nocturnal habits and arranged an additional transmission of the "electronic auras" with a nine-hour phase delay, thus separating the "auras" from the "real bodies" for a considerable period of time. In this particular visual experiment, we not only demonstrate the

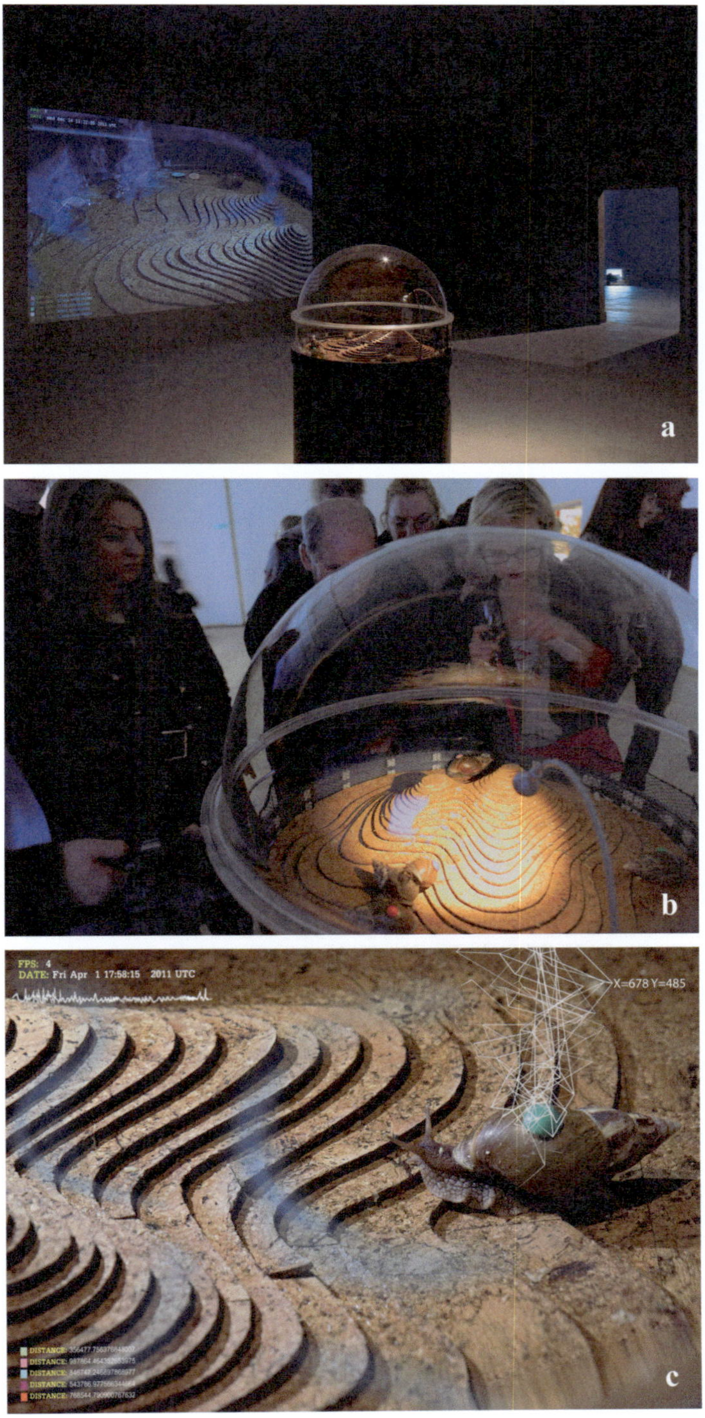

Fig. 4.3 Dmitry Bulatov and Alexey Chebykin. Ongoing since 2011. *That Which Lives in Me*. Bioactive installation with *Achatina fulica* snails and AR-technologies. Supported by the National Centre for Contemporary Art (Baltic branch, Russia) and the Dynasty Foundation for Non-commercial Programs (Moscow, Russia). **a** General View. Image courtesy of the artists, 2012. **b** Installation with visitors. Image courtesy of the artists, 2012. **c** Video still of snail with "electronic aura". Image courtesy of the artists, 2012. **d** Installation detail of snail. Image courtesy of the artists, 2012. **e** Installation detail of snail environment. Image courtesy of the artists, 2012

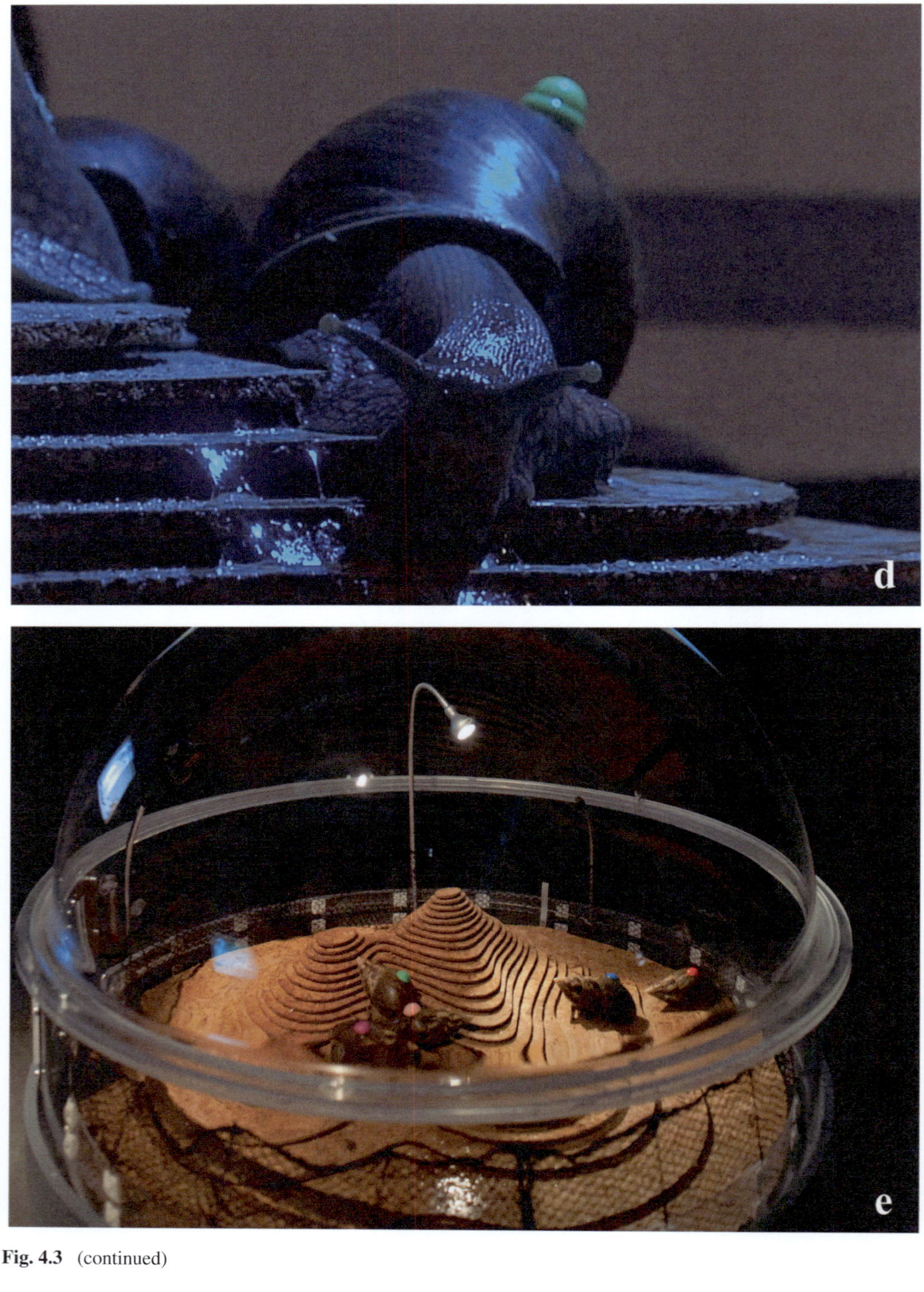

Fig. 4.3 (continued)

abilities of augmented reality to superimpose the real and the virtual in real time, but also to outline the boundaries of applicability for these technologies, liberating spectator perception from the mechanical rhythm of interactivity imposed by the life that surrounds us.

4.3.4 *Formata. Non-terrestrial, Kinetic and Autonomous Agents.* PЯOTO-ALIEИ PЯOJECT, 2020–2023.

By Juan M. Castro

Is there any(body) out there? How do we know that matter isn't self-assembling and evolving in other worlds? Today, at a time when our material culture and anthropogenic impact are expanding beyond Earth, what would it mean ontologically to experience alien material assemblages as active and performative?

FORMATA (Fig. 4.4) is a hybrid installation that stages lively, autonomous blobs inside an experimental reactor emulating the conditions of an extraterrestrial planet with liquid formamide. These vibrant blobs are composed of organic substances similar to those found in meteorites and comets, namely, abiotic fatty acids, amino acids and hydrocarbons. Unlike Earth, the atmosphere of the 'mini-planet' is composed of ammonia, carbon monoxide and argon gases, and its rocky surface is covered with warm pools of formamide. In these pools, the oily blobs move away from stasis: they deform, actively move and self-divide. They become non- equilibrium entities, active material assemblages traversed by flows of matter and energy.

Through the staging of alien active matter, we aim to question the anthropocentric mindset intrinsic to our utilitarian understanding of matter as a passive, inactive or neutral substance. For us it is important that the audience experiences how non-terrestrial entities become operative and autonomous in an alien environment. We aim to reveal how their ability to act is rooted in their materiality, without the need for complex mechanisms, genetic in structions or water. In other words, we aim to exhibit how a self-assembled body, an embedded protometabolism, the capability to deform, move and self-divide are all present within these agents in the form of material agency.

In this work, alienness and the perception that these entities are not yet living but are already more or less alive certainly confronts visitors in multiple ways. This bodily encounter between the audience and alien active matter is critical to our project. It is important not only in the way that these entities help us to materialise the 'other' and the unknown; but also for the way in which they compel us to re-evaluate our place in an active cosmos.

That alien matter can be experienced as active and performative, as a physical and sensual encounter, can not only help us understand how matter itself behaves, but also allow us to rethink the basis for how we perceive and relate to materials, objects and the living. Through these bodily encounters, FORMATA challenges our current relationship with matter, deconstructing the human/non-human hierarchy and replacing it with an "unscaled agency" of living, partially alive and non-living entities.

Website: https://proto-a.jp/

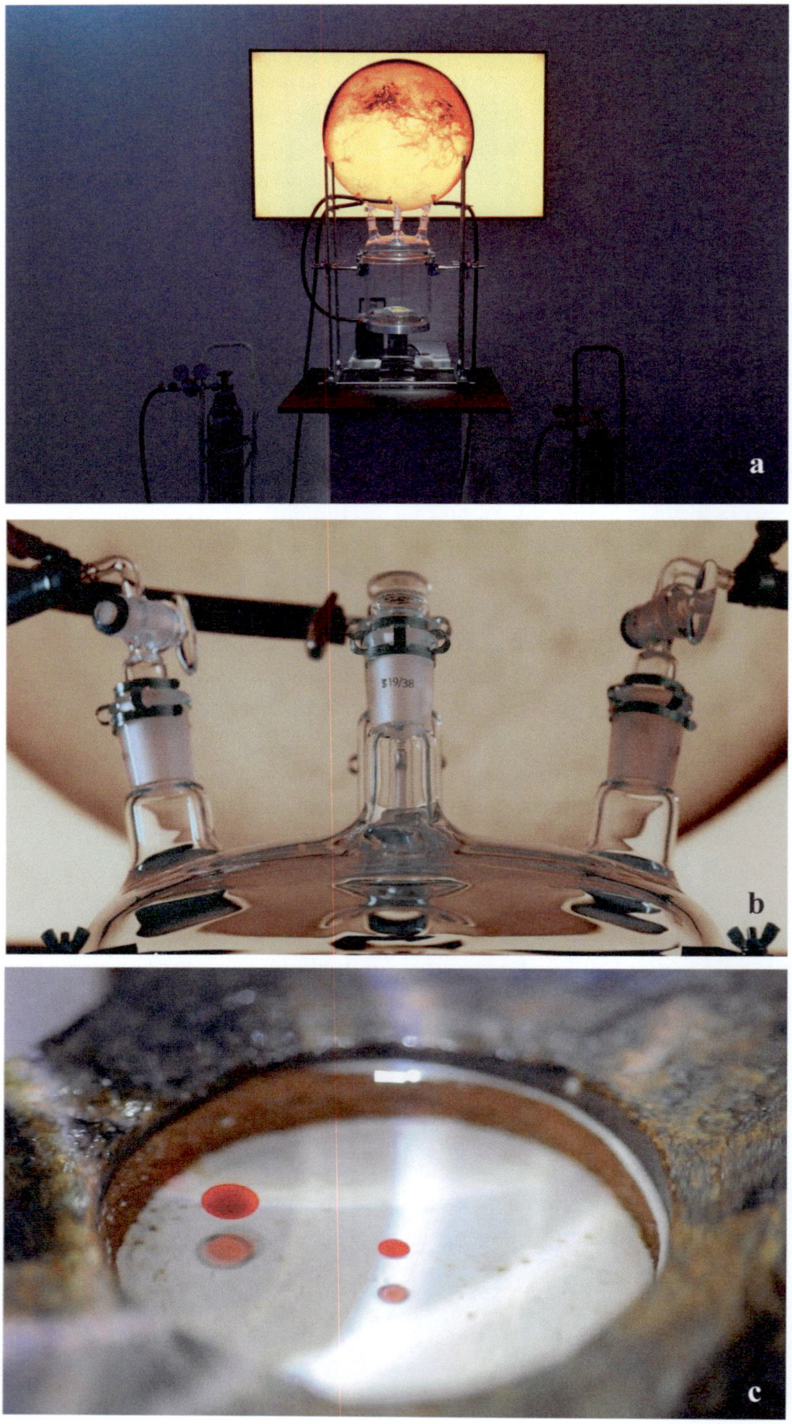

Fig. 4.4 PЯOTO-ALIEИ PЯOJECT. 2021. *Formata. Non-terrestrial, kinetic and autonomous agents*. **a** General view of the Installation, Tokyo. © Juan M. Castro. **b** Image of the reactor's lid and the gas hoses, Tokyo. Image courtesy of Gyre Gallery. **c** Image of the non-terrestrial agents, Tokyo. © Akihiro Kubota. **d** A hypothetical exoplanet covered in seas and lakes of liquid formamide. © Juan M. Castro. **e** Image of the reactor and the rock with formamide pools. 2023 © Juan M. Castro

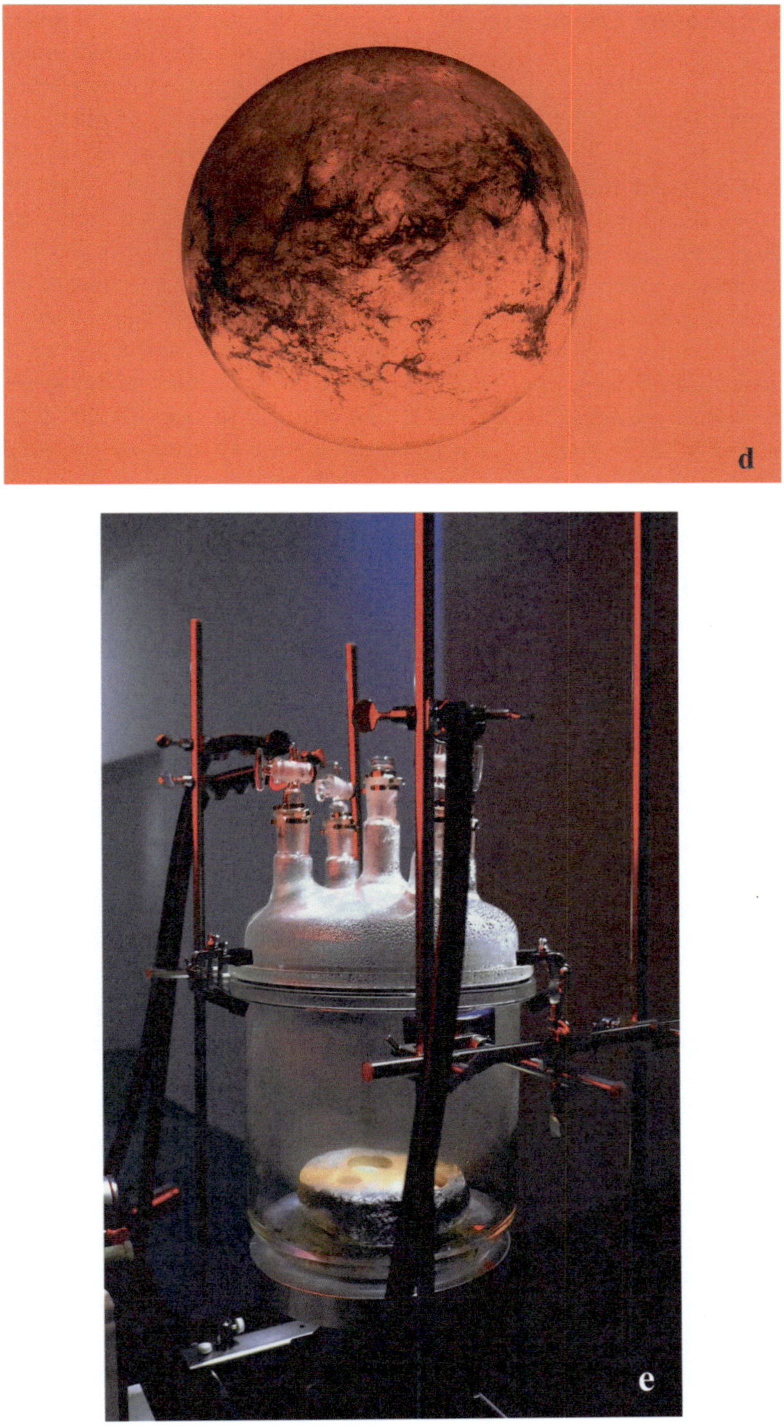

Fig. 4.4 (continued)

4.3.5 *Kolam*. Shanthi Chandrasekar, 2021

By Shanthi Chandrasekar

Kolam (Fig. 4.5) is an art form practised as a daily ritual among Tamil women in India. At sunrise and sunset, they clear the floor in front of their houses and draw a pattern of dots and lines using rice or rock powder as a symbol of welcome and to bring good energy into the household. It is an ancient and living tradition that has been handed down from generation to generation.

Towards the end of 2020, I was approached by Roopal Shah and Sowmya Somnath, to participate in organising a volunteer project involving Kolam and Maths for Washington DC Public Schools. It was planned for January 2021 to coincide with the Tamil harvest festival, (Pongal) and to welcome the incoming Vice-President, Kamala Harris, who is partly of Tamil descent. It was an opportunity to realise an idea that I had thought of years ago and I agreed to design a large Kolam made of 12″ × 12″ tiles, each consisting of a dot in the centre and one of the six basic Kolam shapes, that would be created by the students. The project spurred public interest and we opened the registration to other organisations and communities as well. Roopal and Sowmya organised the project, while I focused on the design, preparation and execution of the Kolam.

Traditional Kolams are an eco-friendly art form, made using natural materials directly on earth and so we chose to use recycled materials such as signboards and cardboards. There are a few core shapes that repeat throughout most Kolams. I realised that each of the six basic shapes could be created using permutations and combinations of two simple templates—a quarter of a circle and a quarter of a square—which could then connect in different ways to create Kolams. The shapes connect with each other at the square nodes, similar to how atoms combine to form molecules, and at an even smaller scale, how a few fundamental particles could create the great many variations in our universe. Due to COVID-19, the participants were asked to print out the templates of the shape assigned to them, design and glue them on the boards based on the guidelines provided to them, minimising any in-person contact required for participation.

I selected a 23-dot Kolam, the notation representing the number of dots in the central, perpendicular line. The Kolam was complex in its twists and turns, yet simple enough to scale up the size, given that it was a fractal, a basic design that could be repeated to create larger and more complex patterns while still retaining its original shape. It also demonstrated multiple symmetries including rotational, translational and scaling—its 90° symmetry made it easier for four teams of volunteers to assemble simultaneously.

Kolams symbolise infinity, as the never-ending line weaves around all the dots, using the shortest possible distance, and gets back to the starting point, forming closed loops. These paths are similar to Euler's circuit, in which a path is traced only once. While there are Kolams that are drawn with a single loop enclosing all the dots, the 2021 Kolam pattern was made up of multiple closed loops to create an intricate network, representing the various communities that participated.

Since it was a site-specific, crowd-sourced project, both the location dimensions and the final number of tiles were vital information for planning. I generated formulas to calculate the number of tiles required and the dimensions of the Kolam by plugging in a variable number of dots for the central line. This process reminded me of the template bank for gravitational waves from coalescing binary black holes or neutron stars, where templates are created using simulations and how the data obtained at LIGO and Virgo is compared with those templates to arrive at the detailed information of the coalescing pair.

The whole project felt like a science experiment as it involved detailed planning and testing of the concept, design and execution. When the site was confirmed, I calculated the layout of the final Kolam pattern, accounting for and including every single tile we had received and created a schedule to assemble it within the allotted four to five hour period. Though at a much smaller scale, this process was reminiscent of the Event

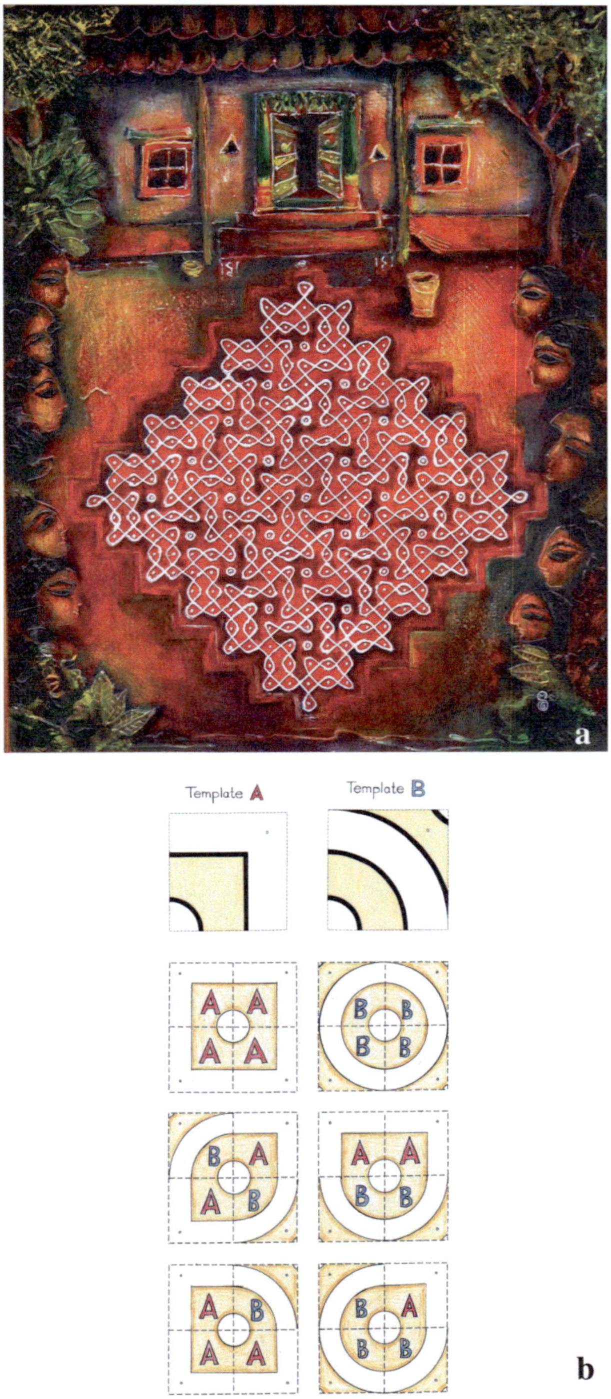

Fig. 4.5 **a** Shanthi Chandrasekar. 2001. *Kolam Celebrations.* Acrylic on Panel. 20″ × 16″, 2001. Photo Credit: Courtesy of the artist. **b** Shanthi Chandrasekar. 2021. *2021 Kolam.* Kolam Templates A & B, and Shapes. Pen & Colored Pencil on Paper. Composite Image, 2020. Photo Credit: Courtesy of the artist. **c** Shanthi Chandrasekar. 2021. *2021 Kolam.* Kolam Trial Run, Maryland, January 2021. Around 800 tiles of 12″ × 12″ each Paper, Mixed Media, Cardboard/Signboard, Glue and Fasteners. Photo Credit: Tony Ventouris Photography. **d** Shanthi Chandrasekar. 2021. *2021 Kolam.* Kolam Final Installation Plan. Pen & Colored Pencil on Graph Paper. Photo Credit: Courtesy of the artist. **e** Shanthi Chandrasekar. 2021. *2021 Kolam.* Kolam Installation, Washington DC, April 2021. 1950 tiles of 12″ × 12″ size. Paper, Mixed Media, Cardboard/Signboard, Glue and Fasteners. Photo Credit: Tony Ventouris Photography

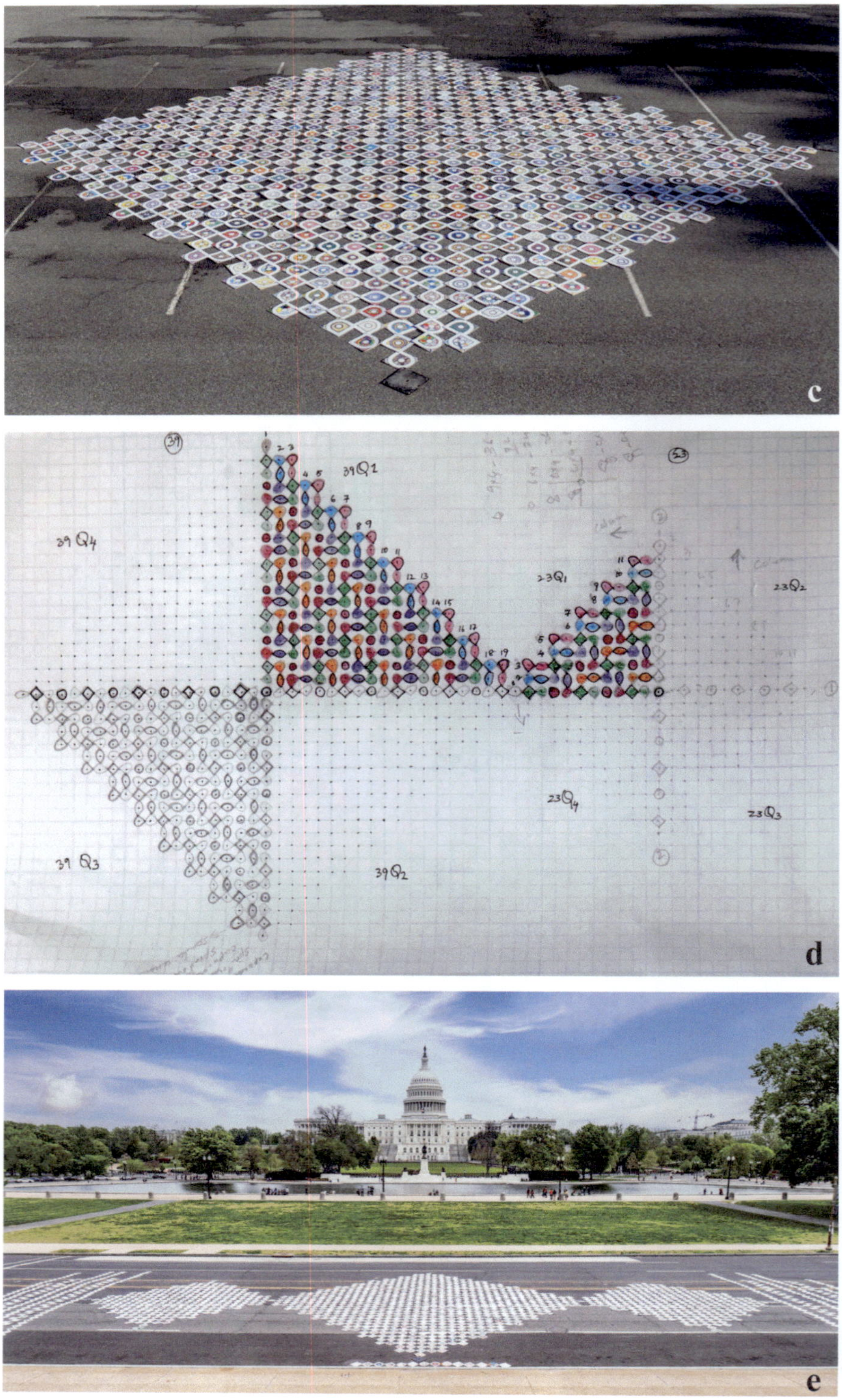

Fig. 4.5 (continued)

Horizon Telescope's photo of a black hole and its shadow using an international network of radio telescopes. All the individual parts had to come together with precision within the limited time, while hoping that external variables like the weather would cooperate.

One word that repeatedly came to my mind during the project was entropy—the whole process was one of alternating states of order and chaos. The ephemeral nature of Kolams serves as a reminder of impermanence, non-attachment, and the cycles of creation and dissolution at various scales. Kolam captures the essence of the cosmos, a small portion of something much larger, and each tile in the installation was the amalgamation of so many individuals' personal narratives, to create something much larger.

Kolams are steeped with philosophy and life lessons, which helped me deal with various setbacks and uncertainties during the project. The dots are said to represent challenges in life, and if one can weave their way around them and get back to the starting point, while still maintaining some kind of symmetry, one would be able to deal with life's ups and downs. Despite the numerous hurdles and challenges, the tiles all fit together perfectly and we were able to assemble the Kolam successfully. We received tiles designed by people between the ages of five to 95 from around the country, including Washington DC Public Schools, a childcare centre and a correctional facility. With the help of family, friends, and volunteers, we assembled a 1950-tile Kolam in front of the Capitol Building in Washington DC in April 2021.

In a Kolam, the grid of dots serves as a scaffold for the lines to weave around, symbolising matter and energy, and in my artistic vocabulary, they denote the particle and wave nature of light. I find similarities between Kolams and the patterns created by vibrations of particles as demonstrated in Cymatics, the shapes of orbitals, or the glial cells in the brain, and these ideas find their way into my art. Leila Haegel, a physicist working on gravitational waves and black holes, and I were partnered during SciArt Initiative's Bridge Science Art Residency, and we discussed how cutting edge science could be interpreted using traditional art forms such as Kolams, Shibori dyeing, or embroidery. I have had similar conversations with Mark Neyrinck, a scientist who researches the cosmic web using origami techniques. Liubov Tupikina, a mathematician and theoretical physicist, and I have also been discussing Kolams and networks analysis and graph theory.

Feynman's diagrams describe the creation and dissolution of virtual particles in the quantum realms and Penrose diagrams describe the other end of the spectrum dealing with space–time, black holes and infinities. Visual tools help in making abstract scientific concepts more approachable, accessible, and easier to communicate. I find Kolams help me understand science and in turn, inspire my art, whether it is a quiet meditative drawing or a community project like the 2021 Kolam.

4.3.6 *Harmonic Dissonance: Brains and Bodies Dancing Together at the Interface of Art and Science.* Matthias Oostrik, Suzan Tunca, Than Van Nispen, and Suzanne Dikker, 2018–2023 (Ongoing)

By Matthias Oostrik, Suzan Tunca, Than van Nispen, and Suzanne Dikker

The description below, including the biographies in Chap. 4, consists of a compilation of texts that have previously appeared in various publications and exhibitions, authored by Matthias Oostrik, Suzan Tunca, Than van Nispen, and Suzanne Dikker. All rights reserved.

Where lies the boundary between our individual and shared identity? How do data generated by our brains, bodies, and algorithms relate to our experience of connectedness? Is there harmony or dissonance between the mind and the body, between us and the people around us, between our sense of connectedness and the data that describe these connections? When we feel in sync with someone, are our brainwaves and bodies literally on the same wavelength?

Harmonic Dissonance (Fig. 4.6) is a series of interdisciplinary works that explore the

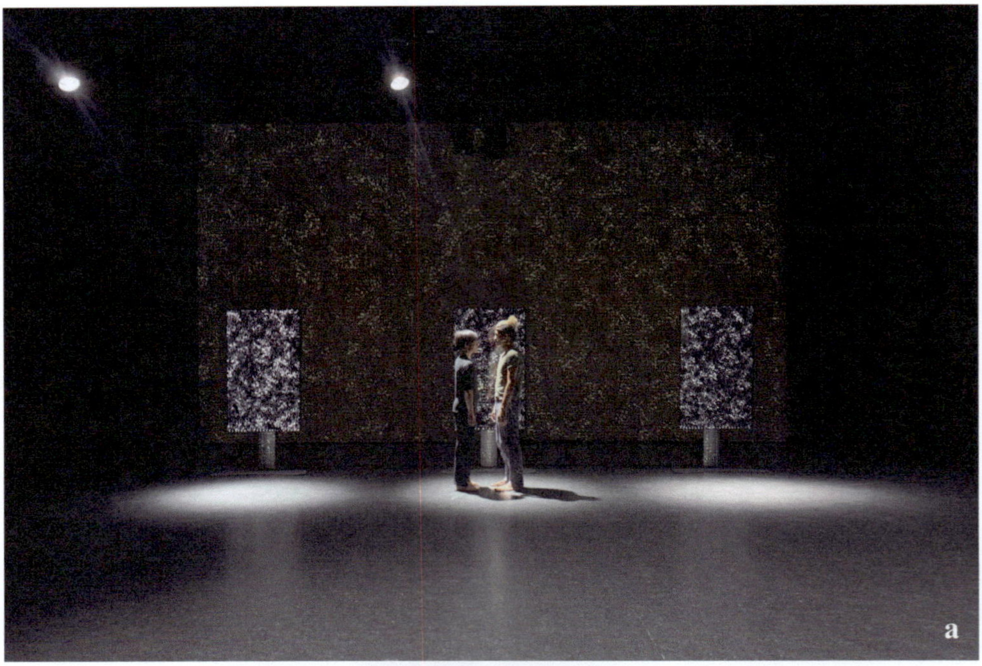
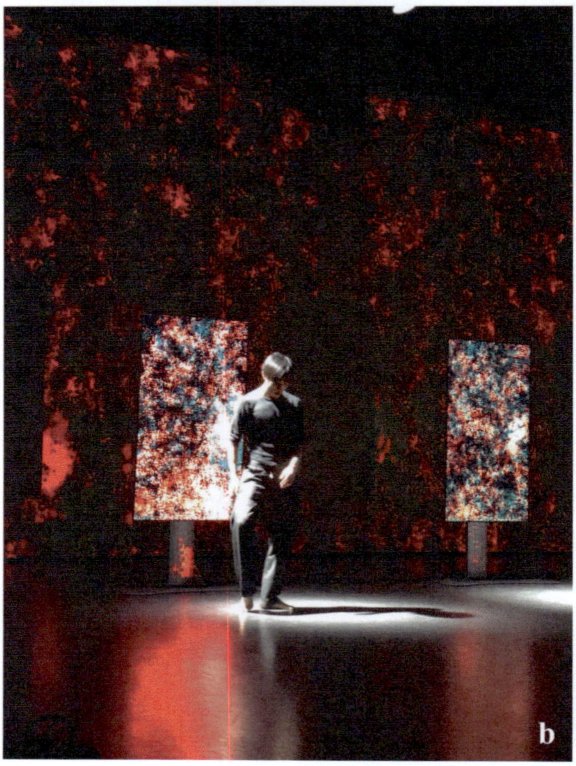

Fig. 4.6 Matthias Oostrik, Suzan Tunca, Than van Nispen, and Suzanne Dikker. 2018–2023 (ongoing). *Harmonic Dissonance: brains and bodies dancing together at the interface of art and science.* **a** ICK Dancers Hiroki Nunogaki and Isaiah Wilson at AHK ID Lab 2020 Amsterdam. Credits: Nim Goede. **b** ICK Dancer Hiroki Nunogaki at AHK ID Lab 2020 Amsterdam. Credits: Nadine Hottenrott. **c** ICK Dancers Beatrice Cardone and Victor Swank Opening Space for Dance Art ICK Dans Amsterdam 2021. Credits picture: Mo Alzoabi. **d** Harmonic Dissonance: HQ. Studio Mellius Amsterdam 2018. Credits: Michel Boulogne. **e** Harmonic Dissonance: Quartet—Doornburgh 2022, credit Michel Boulogne

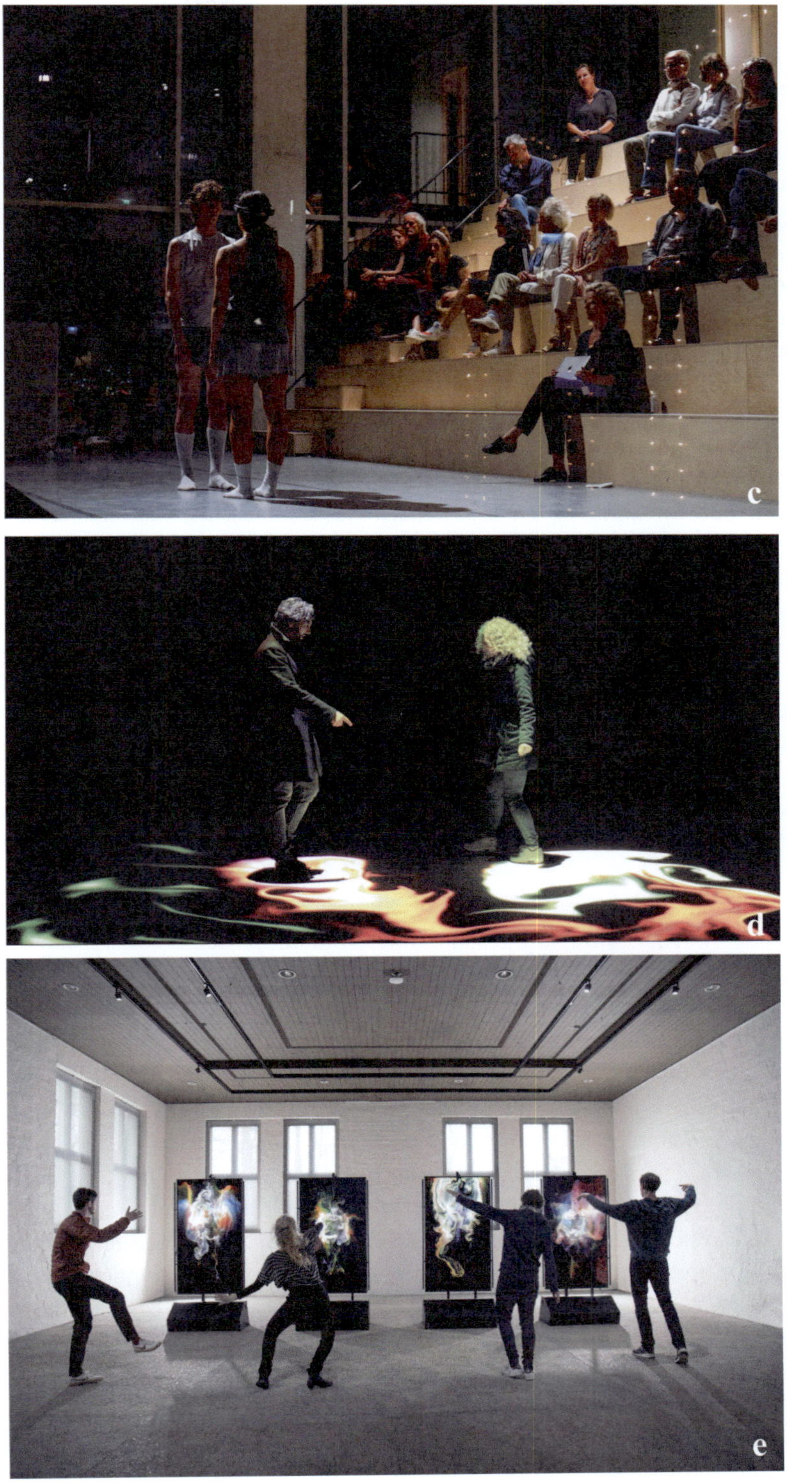

Fig. 4.6 (continued)

measurability of human connectedness from performative, neurocognitive, tech/experiential, and sound design perspectives. In interaction with the audience, Harmonic Dissonance investigates the friction–the harmony and dissonance–between our experience of connectedness and the algorithmic data systems that describe us and our relationships to others.

Harmonic Dissonance aims to expand the boundaries of interactive performance art, which has already extensively explored movement and sound/visuals interactivity, by adding synchrony biofeedback to the performative setting. The project explores how scientifically rigorous experimentation can be integrated into an immersive performative experience; and how an interactive audio-visual environment can translate "objective" data streams (brainwaves, heart rate, movement) into intuitive storylines for performers, viewers, and the general public.

As such, *Harmonic Dissonance* raises questions about virtual and "real" human communication. We are increasingly "connected", yet also increasingly lonely and divided. Can technological interfaces ever substitute face-to-face interactions? The covid-19 pandemic has made us all grateful for the existence of internet-based communication systems, while at the same time bringing to the surface their limitations: mutual gaze and interpersonal synchrony are two core drivers of human connectedness, but in a zoom room you cannot look another person in the eyes and you cannot synchronise with them because of the time delay. By putting human bodies and tech interfaces in the same space, Harmonic Dissonance highlights the disruptive nature of these imperfect systems as well as the immeasurable value and lived dynamics of shared corporeal presence.

Harmonic Dissonance: *Synchron(icit)y* investigates one piece of this larger puzzle: namely how *togetherness* is achieved (in Emio Greco and Pieter Scholten's dance language: "one body", "synchronicity"): How can the intuitive, yet elusive, dichotomy between "togetherness" and "simultaneity" be embodied, conveyed, and measured? Movement and brain activity is recorded from dancers and their audience during 10–15 min performances, which enables the analysis and visualisation/sonification of inter-body and inter-brain synchrony between the dancers, between the audience members, and between dancers and audience. Combining dance and digital interfaces as a sandbox, the project asks how intentionality may trigger "meaningful" (and perceivable) interpersonal synchrony and asynchrony, and various shades on the spectrum between harmony and dissonance.

Harmonic Dissonance: HQ and Harmonic Dissonance: Quartet are playful participatory art-science installations that translate the extent to which body movements and brain activity from its participants become (de)synchronized in real-time. As such, groups of visitors are invited to explore human connectedness as it is mediated by their bodies, brains, and algorithms. In ***Harmonic Dissonance: Quartet***, the audience participates by dancing and moving in front of large audio-visual instruments. Each instrument can be played alone, but the Harmonic Dissonance Quartet is at its best when played collectively. Participants unlock new levels of play by replicating each other's moves, imitating each other's gestures, and synchronizing each other's thoughts. Working together the audience evolves the installation, creating ever-changing compositions that shift between harmony and dissonance.

Art *and* science

A central goal of *Harmonic Dissonance* is to explore how rigorous scientific experimentation, for which replicability and experimental control are key, can be integrated with artistic performance, for which creativity, aesthetics and unpredictability are key. Collecting brain and movement data outside of the laboratory (real-world neuroscience) while people engage in direct, naturalistic social events (hyperscanning: recording brain activity and movement from multiple people simultaneously) provides a unique opportunity to investigate the neurobiological basis of dynamic, naturalistic social behaviour in ways that are impossible in laboratory contexts. For example, the performance structure of *Harmonic Dissonance: Synchron(icit)y* makes it possible to explore detailed research questions that are at the core of the field of interactive neuroscience, including questions pertaining to how synchronous movements and brain activity map onto experiences of simultaneity versus togetherness, or social connectedness.

Participants' movements are video-recorded and analysed with *OpenPose*, a real-time motion tracking software. Additionally, dancers and audience members wear electroencephalogram (EEG) devices that measure their brainwaves. Individual movement patterns, as well the synchrony in movement and brain activity between dancers, dictate real-time audio-visual experiences.

Offline analysis of the data collected helps improve analysis algorithms for both scientific and interactive media purposes. For example, HyPyP (Hyperscanning Python Pipeline; hypyp.readthedocs.io) is an open-source synchrony analysis pipeline for multi-person recordings that establishes analysis standards for the field; and Hybrid Harmony (github.com/RhythmsOfRelating/HybridHarmony) allows scientists and interactive media artists to create interactive synchrony environments.

Documentation and exhibitions

Video documentation & web links

- www.oostrik.net/projects/hdq
- oostrik.net/projects/hds
- vimeo.com/432322711
- vimeo.com/432308273
- www.vimeo.com/266690053
- vimeo.com/312105834
- www.youtube.com/watch?v=lN0b6Y7qz_U
- vimeo.com/607571453

Exhibitions/tryouts/workshops

- Museum De Lakenhal, 2023
- Nxt Museum Amsterdam, 2022
- Buitenplaats Doornburgh, 2022
- ICK Amsterdam Space for Dance Art, 2021/2022
- Studio Knep Warm in Oost, 2020
- IDlab Hogeschool voor de Kunsten Amsterdam, 2020
- Kunsthal Rotterdam, 2019
- Ballet de Marseille (BDM) Fest, 2018
- Buitenplaats Doornburgh, 2018
- Studio Mellius, 2018

Relevant Publications

Dikker S, Montgomery S, Oostrik M and Tunca S (2019) "Using synchrony-based neurofeedback in search of human connectedness". *Brain Art*. Springer, Cham, 2019. 161–206.

Dikker S, Oostrik M, …, and Nijholt A. Artistic Brain-Computer Interfaces: Harmonic Dissonance and The Mutual Wave Machine. *Proceedings of the 2018 CHI Conference on Human Factors in Computing Systems.*

Tunca S and Dikker S Harmonic Dissonance: A case study of experimentation at the intersection of the arts and science. Forthcoming. Experimental Practices: Worlding the Brain.

4.3.7 *Nomadic Plant*, Gilberto Esparza. 2008–2013.

By Jesssica Riskin

These are symbiotic organisms consisting of a robotic system, an organic plant species, and a set of microbial and photovoltaic fuel cells (Fig. 4.7). This is an autonomous species whose metabolic cycle has the potential to repair, on a small scale, some of the ecological damage to its surroundings by recuperating the energy that it takes from the earth. Upon encountering contaminated water, it sucks it up and stores it

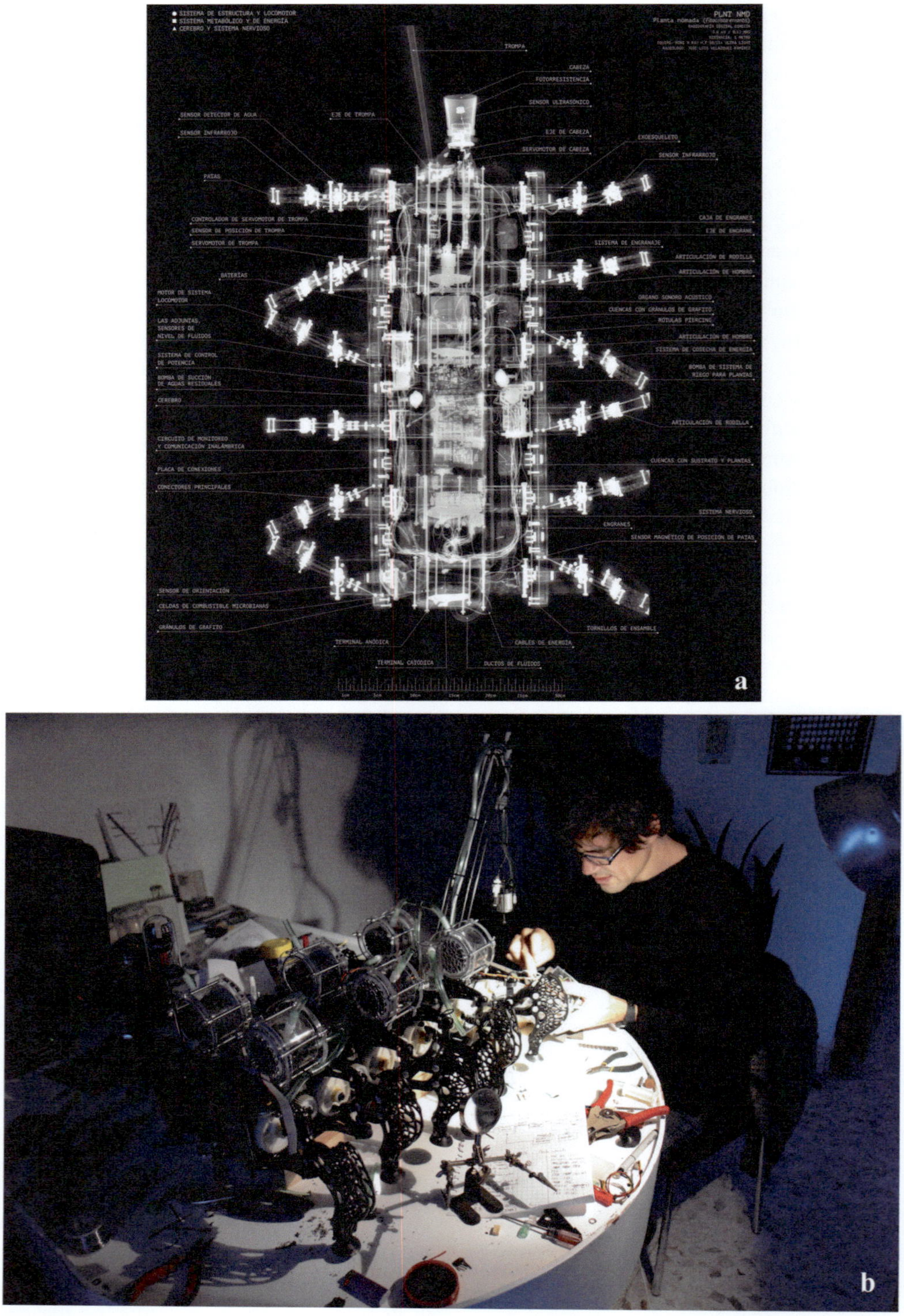

Fig. 4.7 Gilberto Esparza. 2008–2014. *Nomadic Plant (Fitocrista errantis)*. **a** Direct digital radiography. Radiologist: José Luis Velázquez Ramirez. **b** Assembling Nomadic Plants. Image courtesy of the artist. 2008. **c** *Río Santiago*. Image courtesy of the artist. 2009. **d** Encounter with the community. Image courtesy of the artist 2009

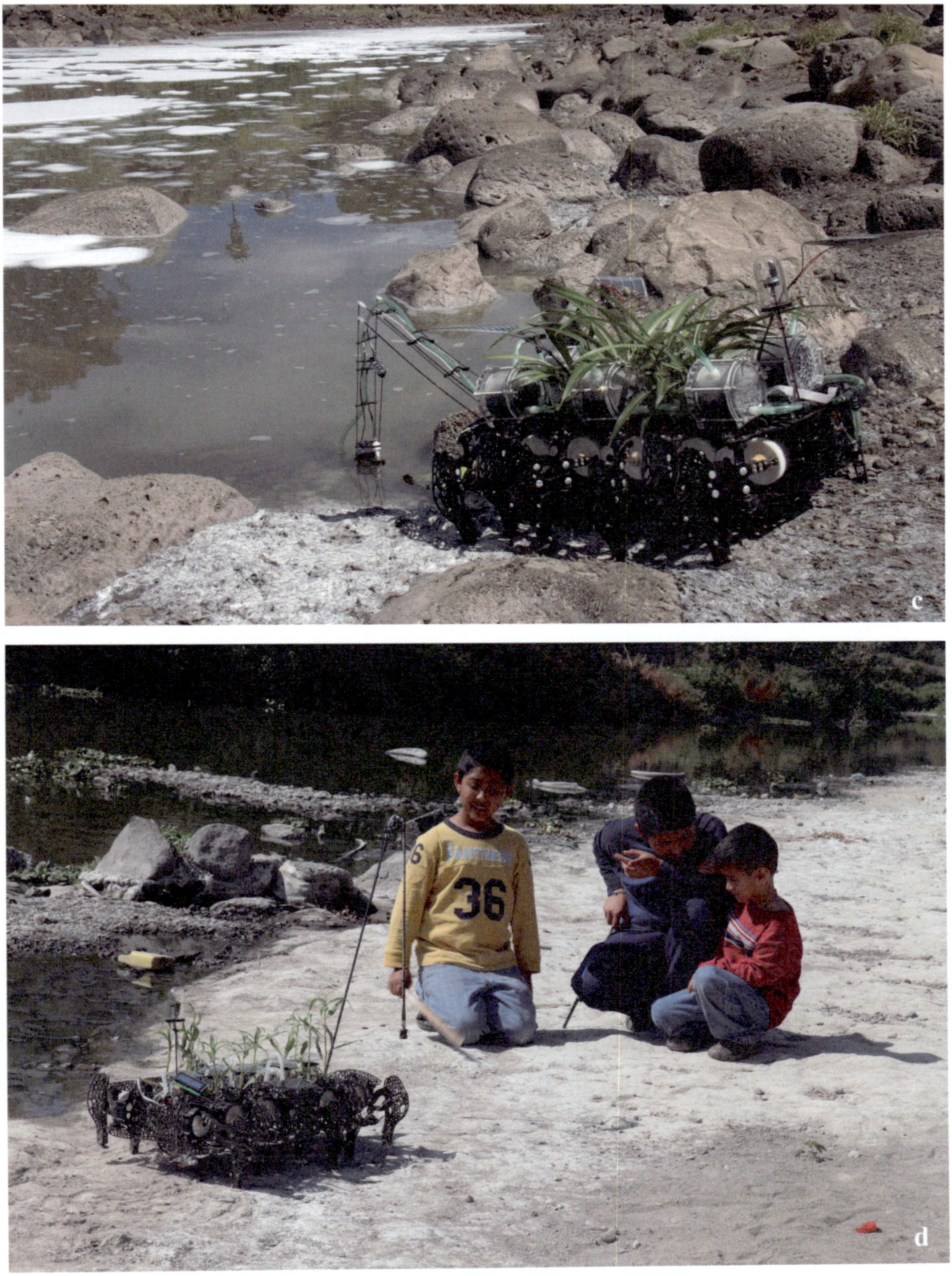

Fig. 4.7 (continued)

in a group of microbial cells, where bacteria and autochthonous microorganisms biodegrade the organic waste and break down toxic substances. This metabolic process generates electricity, which is stored by means of an energy harvesting system in order to charge a set of batteries. The biodegradation process improves the quality of the water, which it provides to the plant species that live in it. The nomadic plant survives in environments that have been affected by water contamination, primarily in areas of ecological disaster affected by industry and the waste products of large urban centres.

4.3.8 *Integrative Illustration of Cells and Viruses.* David Goodsell, 2020

By David S. Goodsell

For the past 30 years, I have created artistic conceptions of the molecular structure of living organisms. These illustrations magnify a portion of a cell and/or virus such that the individual proteins, DNA, RNA, and other molecules can be seen. This is an exciting domain for visualisation since there is currently no experimental method to reveal these views directly. Instead, I integrate information from multiple scientific sources to simulate these nanoscale environments, including biomolecular structures from the Protein Data Bank (www.pdb.org), structures of molecular assemblies from the EMDataResource (emdataresource.org), bioinformatic data from UniProt (uniprot.org), and diverse information from primary reports in the scientific literature such as proteomes and electron micrographs [1].

In response to the Covid-19 pandemic, I have embarked on a series of illustrations to capture the continually-growing state of knowledge on SARS-CoV-2. This work has been presented as part of materials for coronavirus outreach and education at PDB-101, the education portal of the RCSB Protein Data Bank (pdb101.rcsb.org/browse/coronavirus) and all of these images are freely available with a permissive Creative Commons licence to streamline their use.

In the first months of the pandemic, I created an illustration of a generic coronavirus since little was known about SARS-CoV-2 at the time (Fig. 4.8a). This painting is largely based on the virus from the previous 2002–2004 SARS outbreak, since early micrographs of SARS-CoV-2 showed similar characteristics. I also presented a colouring-book version of the image at PDB-101 (pdb101.rcsb.org/learn/coloring-books/coloring-coronavirus), which led to inspiring feedback from parents who used it as a vehicle to discuss the virus and pandemic with their children in a non-threatening way [2].

In the months that followed, I continued to generate updated paintings that incorporated the growing body of knowledge about the virus. This includes a conception of the synthesis and budding of viruses inside a cell (Fig. 4.8b), showing the extensive remodelling of cellular membranes that occurs during the infection. In the Fall of 2020, I held a "CellSpace" workshop that guided dozens of artists in the creation of coronavirus imagery. At the time, much of the discussion in the media was centred on transmission of the virus, so I painted a conception of the virus within a respiratory droplet (Fig. 4.8c). This work and the work from the CellSpace artists were presented in a show "New Ways of Living" in collaboration with SciCommMake2020 (https://www.sigmaxi.org/news/article/2020/07/13/scicommmake-teams-win-funding-to-promote-the-public-understanding-of-covid-19), an event hosted by SigmaXi and ScienceTalk, with the stated goal of exploring "the science of COVID-19, using art to provide friendly, intuitive and evocative windows into the growing body of information about how the virus is transmitted" (ccsb.scripps.edu/new-ways-of-living/).

As we enter the second year of the pandemic, recent work has included a painting of coronavirus fusing with an endosomal membrane, releasing its genetic material into the cell (Fig. 4.8d), and an idealised conception of the mRNA vaccines that are proving essential for reducing, and hopefully stopping, spread of the virus (Fig. 4.8e). In all the work, my goal is to capture the state of knowledge in a way that is both informative and enticing [3, 4], to help put a face on the virus to promote understanding and hopefully action.

Figure 4.8a depicts a coronavirus entering the lungs, surrounded by mucus (green) secreted by respiratory cells, secreted antibodies (yellow) and several small immune system proteins (tan). The virus is enclosed by a membrane (magenta) that includes the characteristic "corona" of spike proteins that will mediate attachment and entry into cells. The nucleoprotein inside includes many copies of the nucleocapsid protein bound to the genomic RNA (purple).

The illustration in Fig. 4.8b shows a cross section through a cell infected with a coronavirus such as SARS-CoV-2. It shows a time point when the virus is actively replicating and new viruses are being created. The cell's molecules are shown in blues and greens, and the viral molecules are shown in pinks and purples. The illustration integrates the current state of knowledge, but many aspects of the virus and its life cycle are still actively being studied, so portions of the painting are speculative.

The painting in Fig. 4.8c shows a cross section through a small respiratory droplet, like the ones that are thought to transmit SARS-CoV-2. The virus is shown in magenta, and the droplet is also filled with molecules that are present in the respiratory tract, including mucins (green), pulmonary surfactant proteins and lipids (blue), and antibodies (tan).

The painting of Fig. 4.8d depicts the fusion of SARS-CoV-2 (magenta) with an endosomal membrane (green), releasing the viral RNA genome into the cell cytoplasm (blue), where it is beginning to be translated by cellular ribosomes to create viral polyproteins. The painting includes speculative elements that are designed to highlight the process, most notably, multiple states of the viral spike protein are shown.

Messenger RNA (mRNA) vaccines developed for the COVID-19 pandemic (Fig. 4.8e) are composed of long strands of RNA (magenta) that encode the SARS-CoV-2 spike surface glycoprotein enclosed in lipids (blue) that deliver the RNA into cells. Several different types of lipids are used, including familiar lipids, cholesterol, ionizable lipids that interact with RNA, and lipids connected to polyethylene glycol chains (green) that help shield the vaccine from the immune system, lengthening its lifetime following administration. In this idealised illustration, the lipids are arranged in a simple circular bilayer that surrounds the mRNA and the PEG strands have both extended and folded conformations. In reality, the structure may be less regular.

References

1. Goodsell D S, Franzen M A and Herman T. (2018) From atoms to cells: Using mesoscale landscapes to construct visual narratives. *J. Mol. Biol.* 430, 3954–3968.
2. Goodsell D S, Voigt M, Zardecki C and Burley S K (2020) Integrative illustration for coronavirus outreach. *PLOS Biol.* 18, e3000815.
3. Goodsell D S (2021) Art as a tool for science. *Nat. Struct. Mol. Biol.* 28, 402–403.
4. Goodsell D S, Olson A J and Forli S (2020) Art and science of the cellular mesoscale. *Trends Biochem. Sci.* 45, 472–483.

4.3.9 *Blood Wars.* Kathy High, 2010–2011

By Kathy High

Blood Wars (Fig. 4.9) is an art and science project that looks at the engagement between two individuals' white blood cells. The premise of

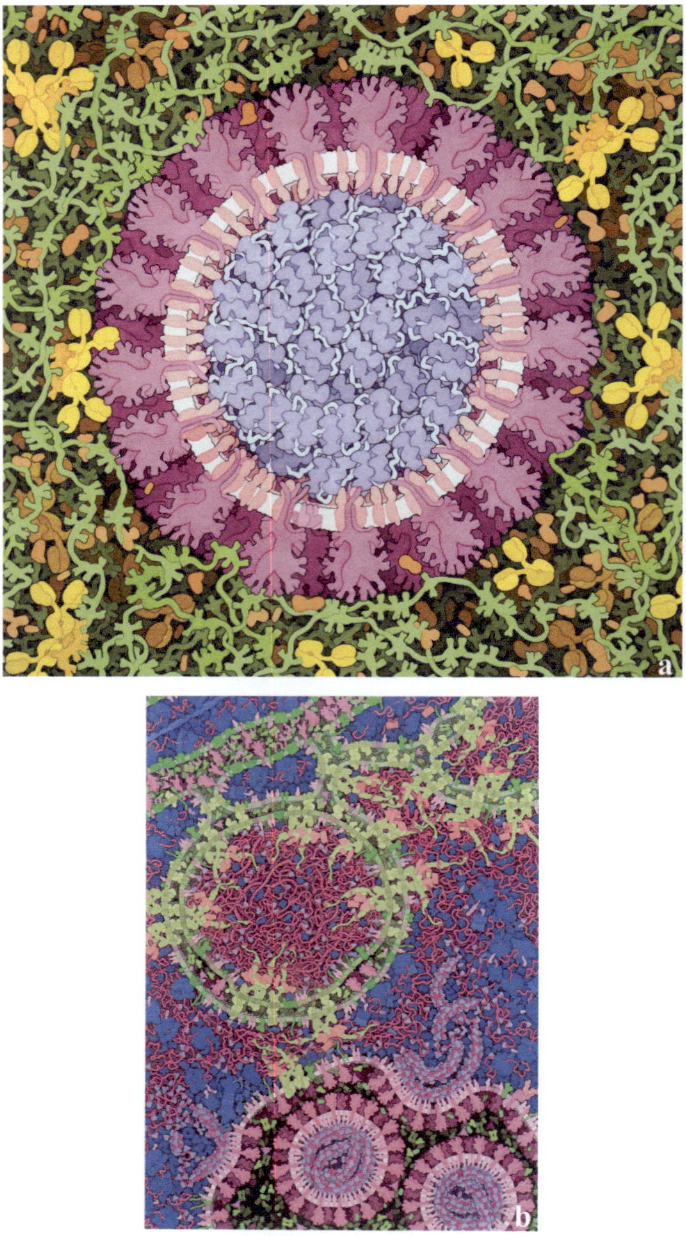

Fig. 4.8 **a** David S. Goodsell. 2020. *Coronavirus.* Painting. RCSB Protein Data Bank; https://doi.org/10.2210/rcsb_pdb/goodsell-gallery-019. Image courtesy of the author. **b** David S. Goodsell. 2020. *Coronavirus Life Cycle.* Painting. RCSB Protein Data Bank; https://doi.org/10.2210/rcsb_pdb/goodsell-gallery-023. Image courtesy of the author. **c** David S. Goodsell. 2020. *Respiratory Droplet.* Painting. RCSB Protein Data Bank; https://doi.org/10.2210/rcsb_pdb/goodsell-gallery-024. Image courtesy of the author. **d** David S. Goodsell. 2020. *SARS-CoV-2 Fusion.* Painting. RCSB Protein Data Bank; https://doi.org/10.2210/rcsb_pdb/goodsell-gallery-026. Image courtesy of the author. **e** David S. Goodsell. 2020. *SARS-CoV-2 mRNA Vaccine.* Painting. RCSB Protein Data Bank; https://doi.org/10.2210/rcsb_pdb/goodsell-gallery-027. Image courtesy of the author

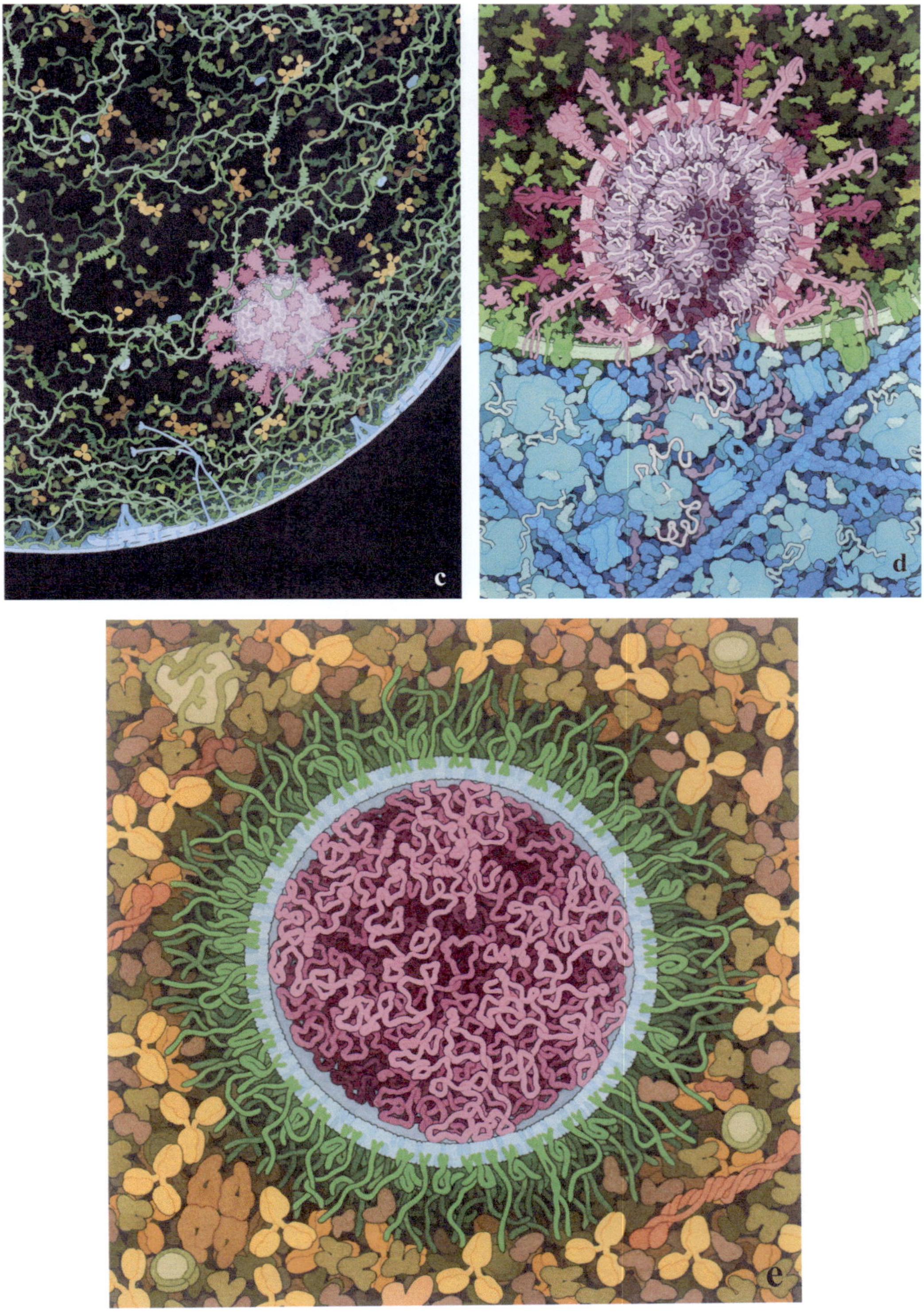

Fig. 4.8 (continued)

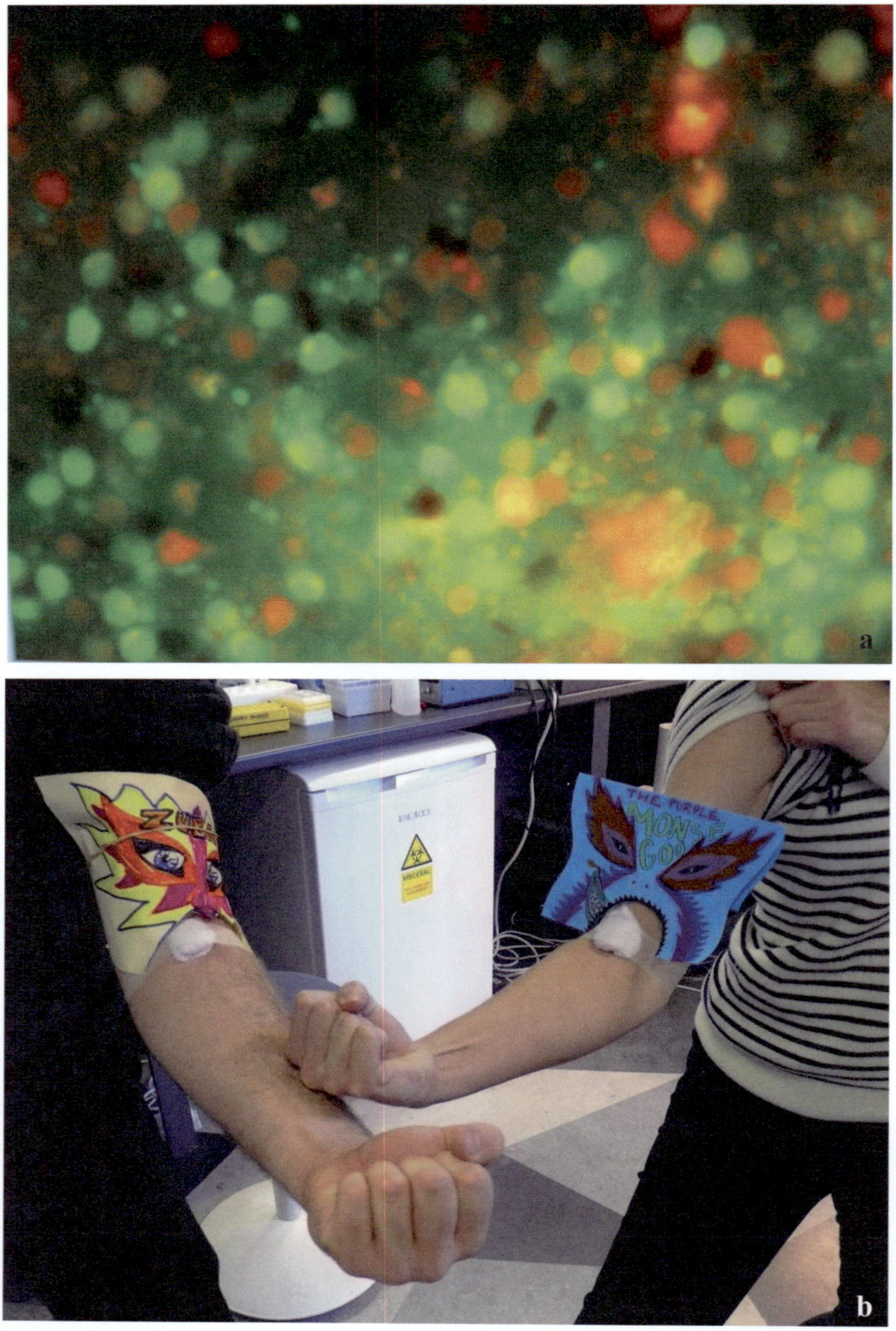

Fig. 4.9 Kathy High. 2010. *Blood Wars*. **a**. *Microscopic image of two participants' competing white blood cells (stained green and orange)*. SymbioticA. Image courtesy of the artist, 2010. **b** *Fierce competitors in Blood Wars competition*. Science Gallery Dublin. Visceral exhibition. SymbioticA. Image courtesy of the artist, 2011. **c** ***Blood Wars logo*** Separation of red and white blood cells from human blood sample. SymbioticA. Courtesy of the artist, 2010. **d** *Blood Wars logo*. Courtesy of the artist, 2010. ***Separation of red and white blood cells from human blood sample***. **e** *Human blood cell separation process*. SymbioticA. Image courtesy of the artist, 2010

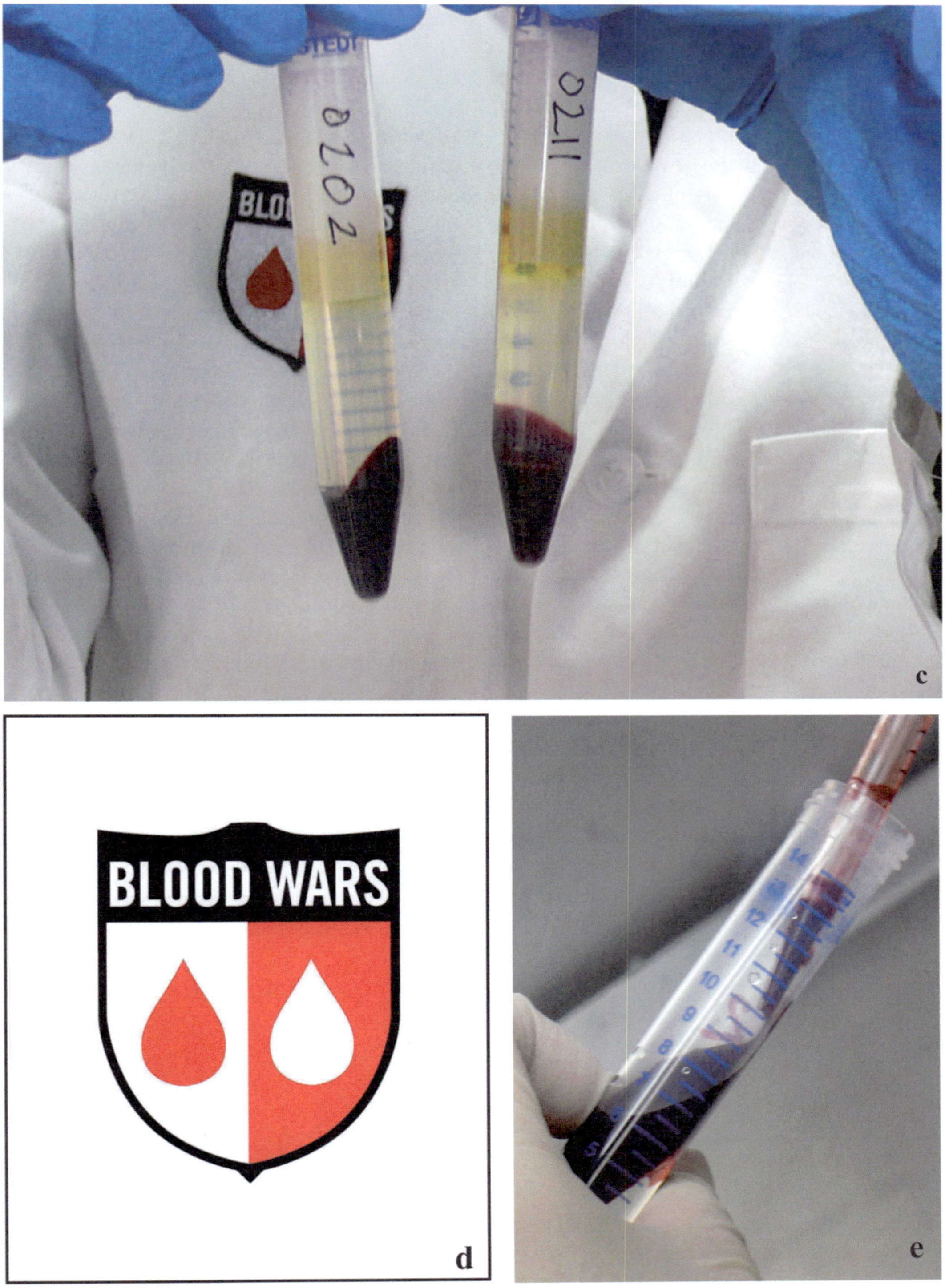

Fig. 4.9 (continued)

the project is to investigate the immune system, among other things. *Blood Wars* takes place as a tournament where matches are played –similar to the World Cup, resulting in an ultimate victor. Who has the strongest blood?

Blood Wars is a two-part project: a live performance of blood-letting/blood collection; as well as a gallery installation of resulting microscopy films and a mapping of the results.

Blood Wars is a real tournament between different individuals' white blood cells as they vie for dominance in the petri dish. Using human blood samples, drawn by professional phlebotomists, *Blood Wars* creates a series of encounters where the cellular 'winner' of each match goes on to fight the next participant. *Blood Wars* engages in blood duels—and the resulting data is collected.

Blood Wars tournaments playfully push the idea of combating opponents and pose questions about the body's 'defence' system—or rather how we think about our 'immune' system. The games play in the moment and highlight the immediate struggles of the cells. But what defines 'strength' or success in this case? These *Blood Wars* tournaments raise more questions than they resolve.

Blood Wars was researched and developed 2009 and 2010 at SymbioticA, University of Western Australia. The project was funded by a 2010 Fellowship from the John Simon Guggenheim Memorial Foundation with additional support from Rensselaer Polytechnic Institute, NY, USA.

4.3.10 *Impulse*. Yunchul Kim, 2018

By Yunchul Kim

Drawing inspiration from the intricate networks of a tree's circulatory system, *Impulse* (Fig. 4.10) implements a complex mechanism that propels fluids in and out, like the flow of sap. This coordinated movement occurs within three water reservoirs nestled inside the core, facilitated by the action of non-pulsating pumps. The conduits enabling this hydraulic action are sensitive to the interaction of water velocity and air bubbles, causing the connecting tubes to resonate. This resonance elicits tremors that reverberate throughout the entire structure. The various aqueous substances trapped within the reservoirs appear to surge and ripple, a dynamic effect further enhanced by the competitive choreography of the pumps, following the strategic dance of computer algorithms generated in real-time.

As visitors engage with *Impulse*, they encounter an artificially animated entity through the visual spectacle of fluid motion and the auditory tapestry created by the device's sound waves. *Impulse* is originally exhibited in conjunction with a work called *Tubular*, which sends micro-hydrodynamic bubbles and fluids into the interior of *Tubular*, causing it to change optically by changing its material refractive index, which then extends into the optical realm.

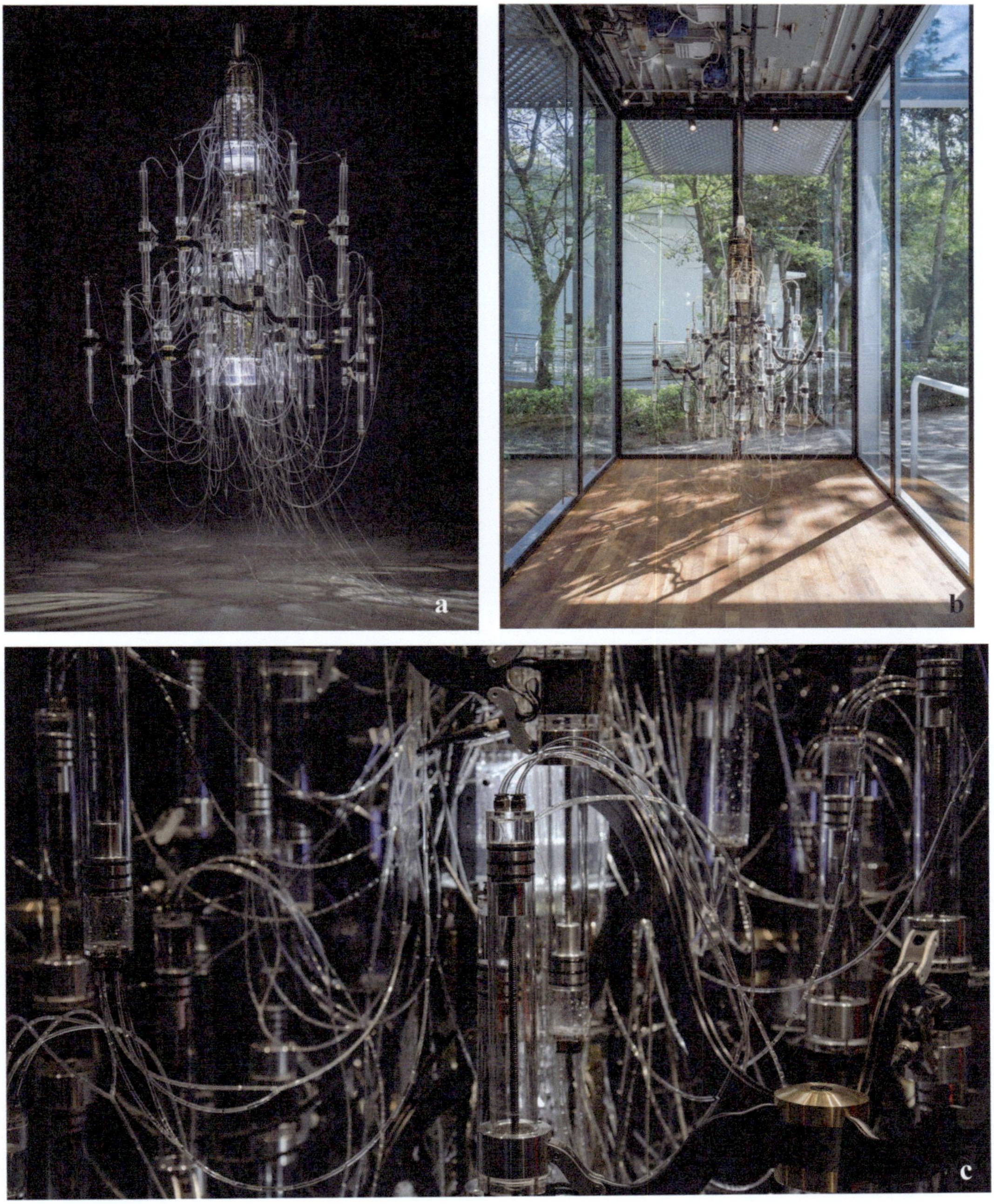

Fig. 4.10 Yunchul Kim. 2018. *Impulse*. Non-pulsating pump, solenoid valve, micro controller, acrylic, aluminium. 230 × 200 cm. **a b** General view. Photo by Mark Blower (a) and Roman März (b). **c** Detail. Photo by Studio Locus Solus. **d** Detail. Photo by Rob Battersby. **e** Detail. Photo by Rob Battersby

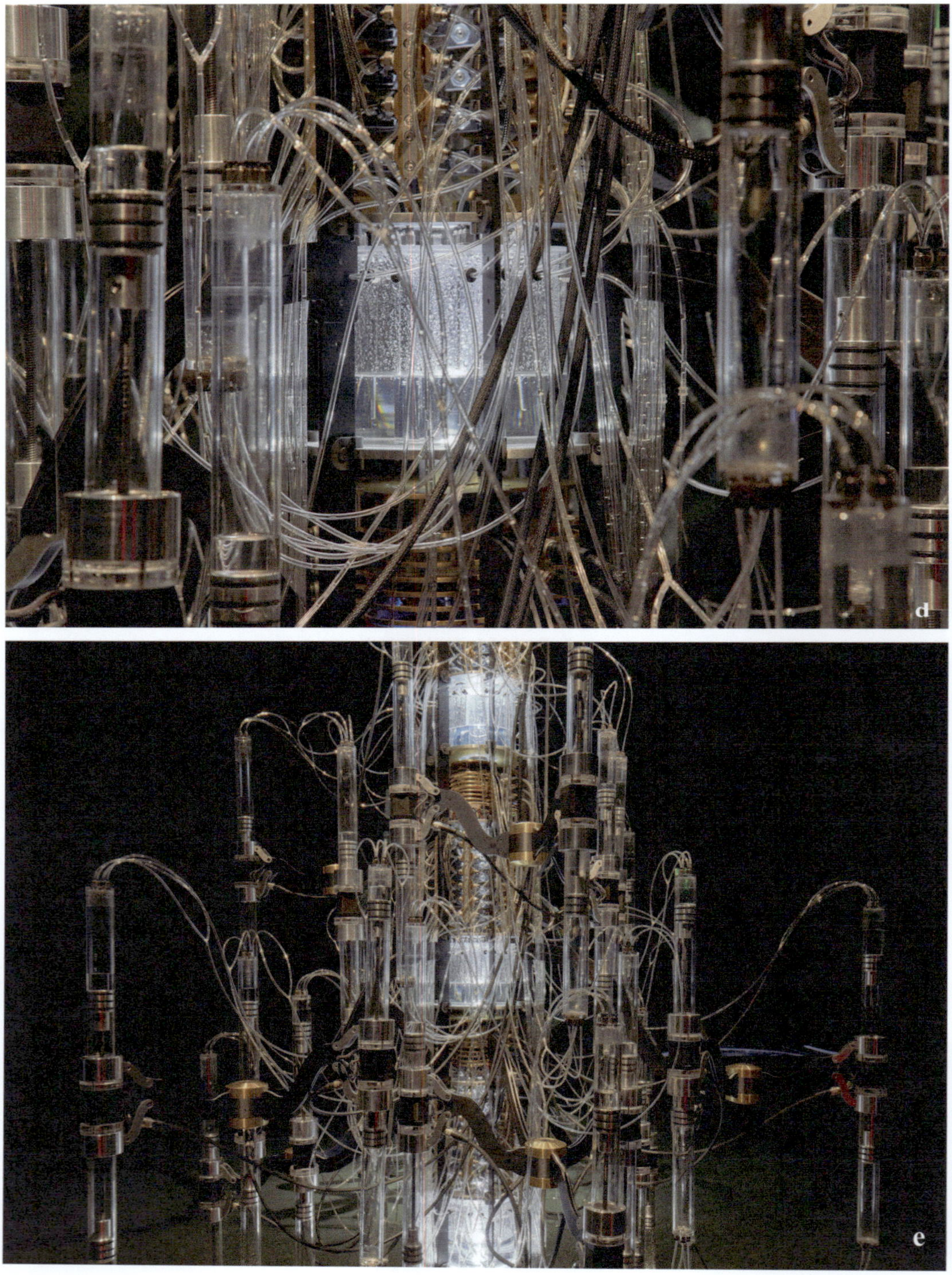

Fig. 4.10 (continued)

Nevertheless, *Impulse* can be exhibited alone, and the inverted tree form of the artwork transforms into a device that allows us to think beyond machines and devices, into a symptom generator that, like Gaston Bachelard's Water and Dreams, presents us not with images, but with matter and symptoms in a pulsating reaction with our bodies.

4.3.11 *Microbiome Rebirth Incubator.* François-Joseph Lapointe and Marianne Cloutier, 2016–2023

By François-Joseph Lapointe and Marianne Cloutier

It is already known that babies born naturally pick up a lot of vaginal (and some faecal) bacteria as they go through the birth canal, whereas babies born by Caesarean section (C-section) miss out on this crucial mix of beneficial microbes. This difference in delivery modes is still important at later ages as C-section increases the risks of allergies, autoimmune diseases and obesity. One way to circumvent this problem is to seed the new-born with the natural microbiome of the mother, by placing a sterile cloth in the vagina one hour prior to a planned Caesarean and wipe the baby's mouth, eyes, face and skin immediately after birth. Unfortunately, in cases of emergency C-sections vaginal microbiome seeding is not possible.

The *Microbiome Rebirth Incubator* (Fig. 4.11) is designed to address this important issue and to investigate the possibilities of repairing or transforming the microbiome of a new-born. The Incubator is a life-size glass container, large enough to fit a baby. It is filled with amniotic fluids (or physiological serum) and connected by a transfusion tube to an Erlenmeyer containing bacteria isolated from the vaginal microbiome harvested from a collaborating mother-donator. Breastmilk is also added to the mix to provide inoculation of beneficial bacteria. A functioning pump ensures the movement of the fluid, and the circulation of air into the device. A faucet at the bottom of the incubator allows to collect the liquid in order to seed new-born babies with a cocktail of bacteria selected to restore their microbiome as if they were just out of their mother's womb.

Beyond investigating biology and health perspectives regarding the healing powers of the microbiome, this artwork also seeks to address a feminist issue, the often traumatic experience of an emergency C-section. Indeed, for many women, Caesarean delivery is experienced as a violent takeover of the medical body on their own pregnant body, a loss of agency of the unique and precious moment of giving birth. By considering the possibility of erasing the microbiome resulting from a C-section and replacing it with this hybrid microbiome derived from the biological fluids of the mother, this artwork is exploring the idea of a symbolic erasure of the medical power over the woman's body, as well as the possibility of a re-empowerment of the mother and her child.

The artwork is composed of the incubator, a soundtrack where six women narrate their emergency C-section stories, a performance during which the incubator is seeded with vaginal bacteria and breastmilk, and a documentation video.

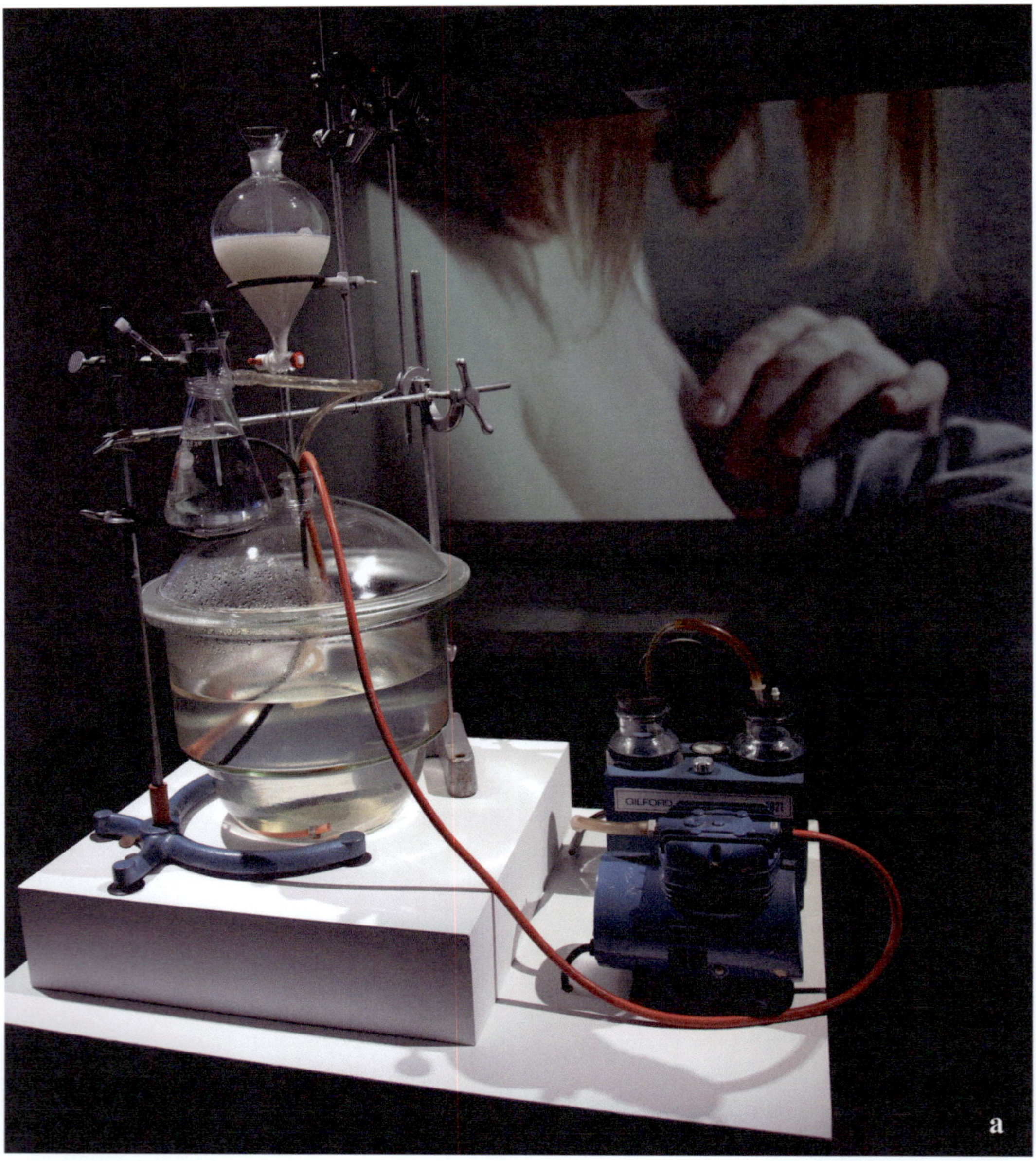

Fig. 4.11 François-Joseph Lapointe and Marianne Cloutier. 2016–2023. *Microbiome Rebirth Incubator*. **a** Installation of the Microbiome Rebirth Incubator filled with physiological serum, vaginal microbiome and breastmilk, alongside a video of the seeding ritual, as part of the exhibition Monstra te esse Matrem, Science Gallery London. Credits Mathieu Girard/Centre Bang, 2023 **b** The seeding ritual of the Microbiome Rebirth Incubator during which the different containers are filled with vaginal bacteria and human breastmilk. This performance was repeated on a daily basis as part of the Taking Care exhibition. Performance during Ars Electronica opening. Credits Vog.photo/ Ars Electronica, 2018. **c** General view of Installation of the Microbiome Rebirth Incubator filled with body fluids for the Spare Parts exhibition. Credits Richard Eaton/ Science Gallery London, 2019

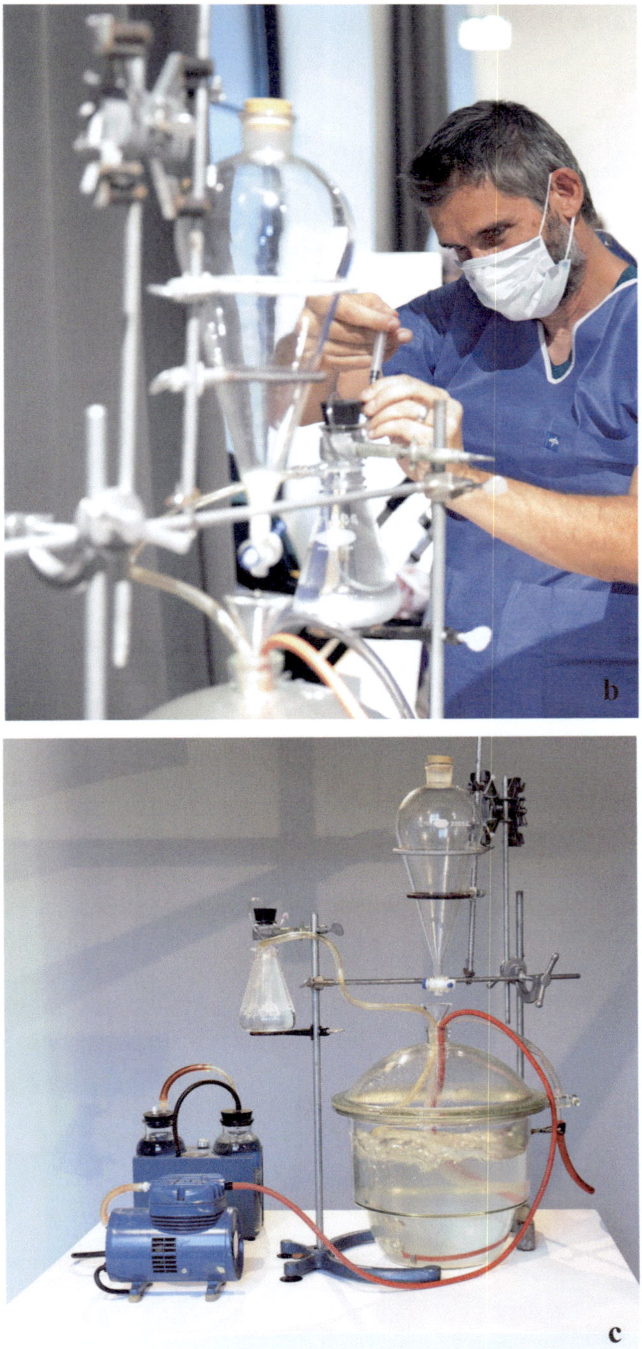

Fig. 4.11 (continued)

4.3.12 *Plastic Is Land*. Max Liboiron, 2018

By Max Liboiron

> This machine here is made from mother earth. It has a spirit of its own. This spirit probably hasn't been recognized and given the right respect that it should. When we work in a world of automated things, we forget that... everything is sacred, and that includes what we make.
>
> Shawn Wilson, *Research is Ceremony* (2008).

For my day job, I'm a marine plastic pollution scientist. I spend hours, days, and years intimately connecting with plastics. *Plastic is Land* (Fig. 4.12) are microscopic plastic landscapes encountered during this laboratory work, specifically during a plastic ingestion study on Eider ducks and Arctic char gathered in a subarctic region of Labrador called Nunatsiavut. Photographs are of scientific samples of gastro-intestinal tracts of animals used in plastic ingestion monitoring, which form micro-scale environments under the dissecting microscope. These images are a combination of found and staged vignettes among samples from the CLEAR laboratory at Memorial University in Canada.

The photos are an effort to counter dominant, activist narratives about plastic pollution as a scrounge, a foreign invasion, a force to be annihilated. Plastics are made from natural gas and oil that began as ancient plants and animals, our relatives. They've been extracted from the ground and put into cracking towers that force them into polymer relationships that are against their nature. They've been violently moved from one set of relations to another.

Yet even when twisted into these new relations, plastics remain part of land relations and living things. In fact, one of the hardest and most time-consuming parts of laboratory studies of plastic ingestion is telling plastics apart from non-plastic look-alikes & act-alikes. Is it a rock or a plastic? A bit of coral or a plastic? Telling the difference takes time and expertise. In the photos, some of the plastics are obvious, but some are not. Some of the items a viewer might assume to be plastic are mussel shells or bone. The point is not to sort out plastics from non-plastics definitively and forever–plastic pollution does not offer such a purity model!—but to understand how the sorting will never be complete and will never be certain. Plastics are land.

This isn't to say that plastics are wonderful, and we should all get along, but that Land is made of people, events, memories, animals, plants, air, soil, water, and since they were mass produced after 1945, plastics. These are our collaborators for land relations moving forward. We want to be healthy and flourish and putting a stop to the continued production of plastic pollution that twists ancient organics into unnatural relations is part of that. The other part is to live with plastic-and-land without stigma or appeals to purity.

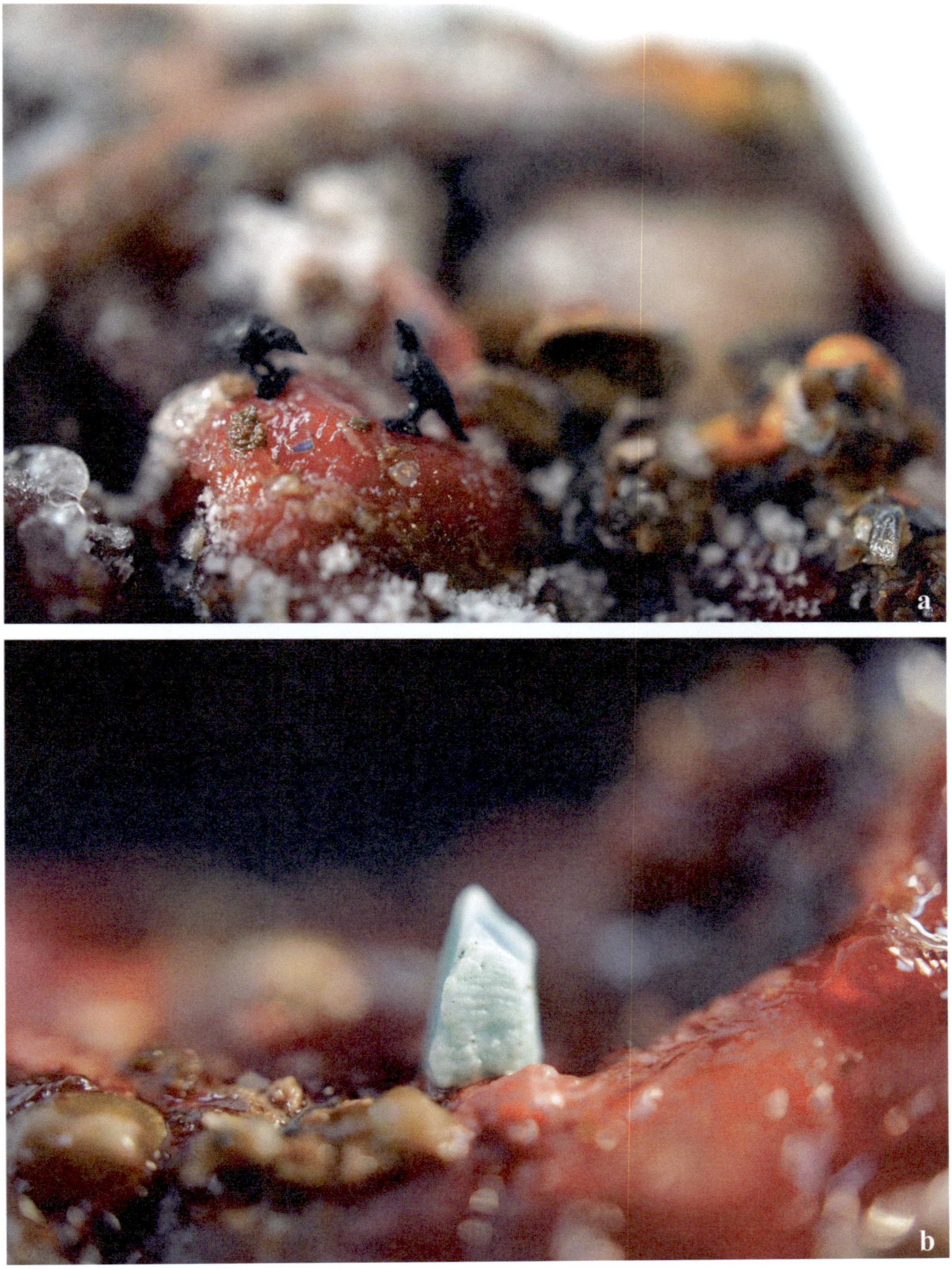

Fig. 4.12 Max Liboiron. 2018. *Plastic Is Land (Eider Ducks and Arctic Char)*. **a** Muscle crows. Digital photograph. Images courtesy of the author. **b** Turquoise fragment portrait. Digital photograph. Images courtesy of the author. **c** Green fragment background. Digital photograph. Images courtesy of the author. **d** Yellow fragment crow. Digital photograph. Image courtesy of the author. **e** Blood two birds. Digital photograph. Images courtesy of the author

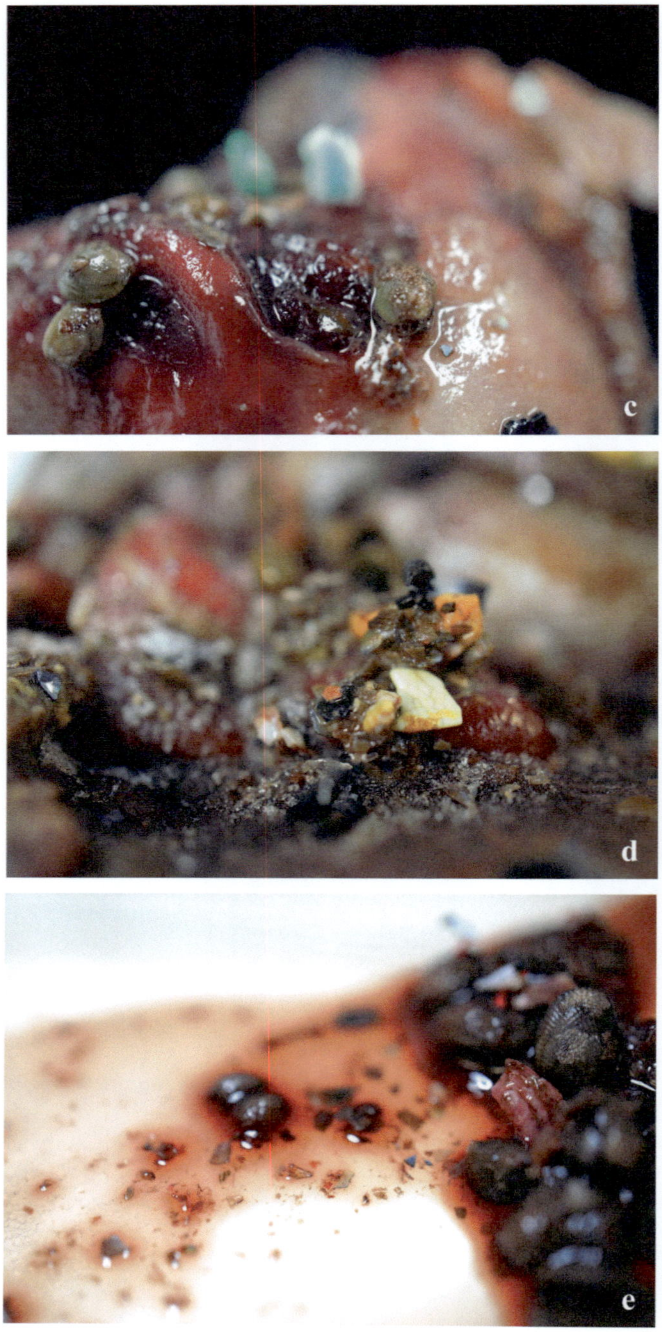

Fig. 4.12 (continued)

4.3.13 Primate Cinema: Apes as Family. Rachel Mayeri, 2011

By Rachel Mayeri

It has been many years since I completed *Apes as Family*. It remains the most ambitious, improbable and collaborative work I have ever been involved in. The project (Fig. 4.13), in its simplest form, consisted of making a film expressly for an audience of chimpanzees, and recording the chimpanzees watching the film. Chimpanzees in captivity are shown a video as a form of enrichment, but no artist had ever made a film for them. The idea, so exciting at first, was to consider other minds and perhaps reflect those minds in a way understandable to primates through filmmaking.

To talk about the process there are several threads, the one about research and another one that has to do with increasing access to support for the creation of art and science. *Apes as Family* is the result of many years of engagement as an artist with primates and primatologists. First *Baboons as Friends*, which was a translation of a sexual consort of baboons as a film noir movie. For that project, I worked with a primatologist who became a long-time collaborator, Deborah Forster. We conducted a workshop on how to act like an animal, with actors and non-actors learning about primate sociality and movement. I attended many primatology congresses and participated in many exhibitions. Later, I worked with baboons and squirrel monkeys and with keepers and scientists at a breeding centre in France. At that time, I was able to propose to the Wellcome Trust the project of making chimpanzee films with Arts Catalyst. My interest in primates had evolved from media depictions to cross-species comparisons with the minds of nonhuman animals.

The idea was perhaps inspired by an article (and later a book) written by Charles Siebert, *Planet of the Retired Chimps* (2005), who met a chimpanzee in a nursing home whose favourite TV show was "Emergency Room." I wondered if the stretchers or lab coats were attractive because many of these chimpanzees had been raised in laboratories. My first thought was to make a hospital drama for chimpanzees. But the psychologist I worked with at Edinburgh Zoo wisely dissuaded me from touching any morbid material: zoos, especially in the UK, are very sensitive to negative media. Beyond that first idea, I wanted to know what the chimpanzees liked to see. I asked many zookeepers for their stories. I researched all the articles on chimpanzee enrichment. I approached the problem of how to make chimpanzee films as a market researcher. In fact, I called one to ask if the process could be applied to chimpanzees; the answer was no, because they cannot be interviewed.

I had a week to test the material with the chimpanzees. I brought in actors from my workshop on how to act like chimpanzees, and we performed all the dramas of chimpanzee life: demonstrations of dominance, sex, eating, wearing masks, doctor's visits. I shot a video of the chimpanzees at the Los Angeles Zoo. I collected material: cooking shows, wildlife films, cartoons, drums, commercials with an animatronic chimpanzee. We placed the images on two monitors in the Edinburgh Zoo research room, with the idea that we could determine the preference for which monitor the chimpanzees spent more time with. Of course, at first the chimpanzees were very excited about the presence of a new group of researchers. What kind of hurdles would they have to jump through? Maybe there would be food! After the initial hazing (they showed off and spit on me), the individuals settled into a few videos, really expressing their interest at times, sticking their heads very close to the screen or, in one case, making everyone in the pod show off: hooting, jumping, with erect hair. Those were the images of the chimpanzees showing off and screaming at the Los Angeles Zoo.

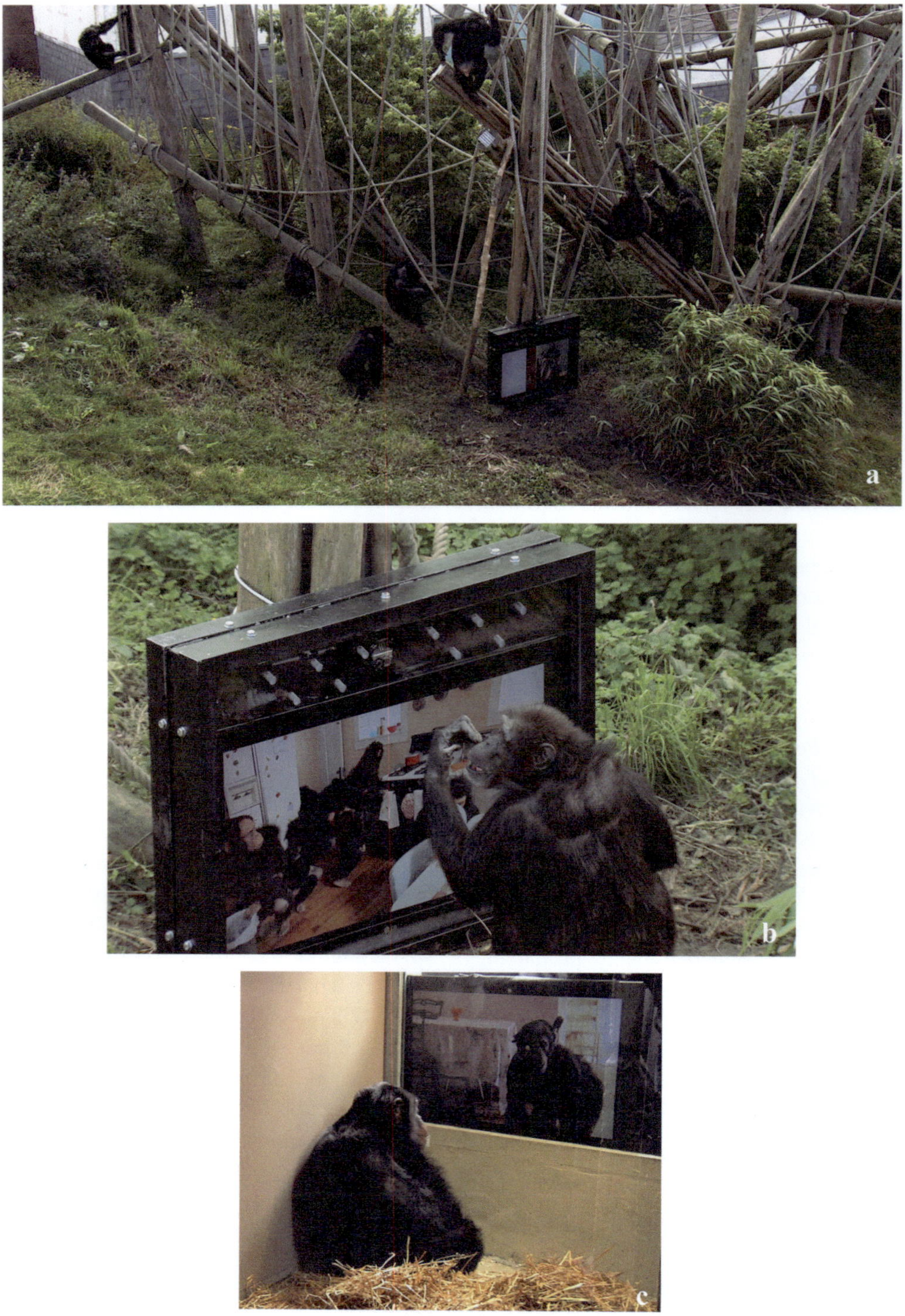

Fig. 4.13 Rachel Mayeri. 2011. *Apes as Family*. **a** Edinburgh Zoo chimpanzees encounter a monitor in their enclosure for the first time. Liverpool. Photo credit: Rachel Mayeri. **b** A chimpanzee investigates the monitor in her enclosure. Liverpool. Photo credit: Rachel Mayeri. **c** A chimpanzee watches Apes as Family in the research pod at the Edinburgh Zoo. Liverpool. Photo credit: Rachel Mayeri. **d** Two screens show the film for chimps and their reaction in an installation at the Edinburgh Art Festival. Edinburgh. Photo credit: Rachel Mayeri. **e** Director Rachel Mayeri and actors in chimp suits on the set of Apes as Family. Liverpool. Photo credit: Matt Chaney

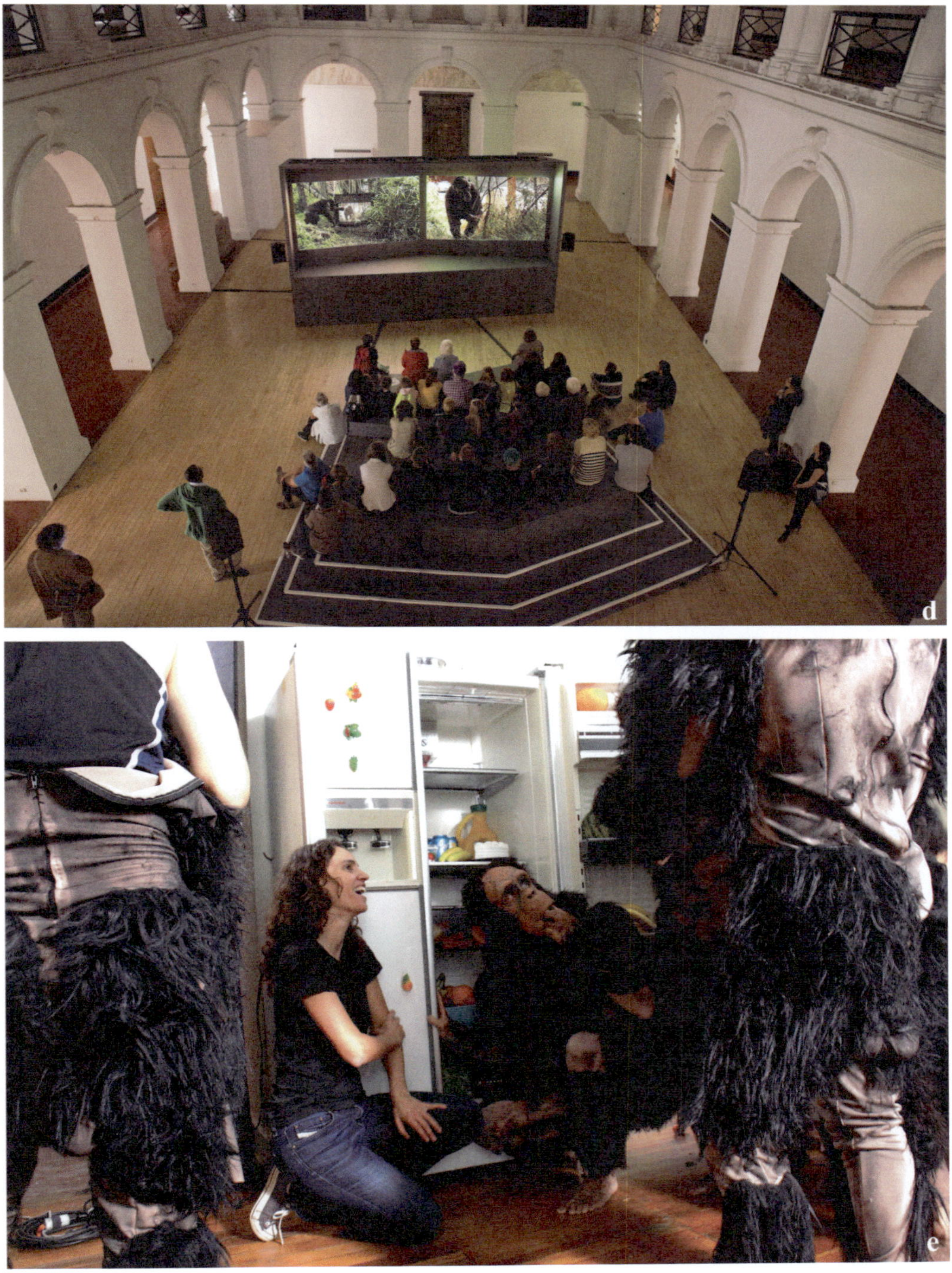

Fig. 4.13 (continued)

The script and production were stressful. I needed to create a story with a cast of chimpanzees, and of course that was impossible. Using real chimpanzees as actors is unethical. 3D animation is prohibitively expensive. We were able to indulge in a day with an actor in an animatronic chimpanzee costume, whose facial expressions are controlled in part by two puppeteers. One chimp alone would not allow any social situations, so my cast would be one "real" chimp and several fake chimps. The "real" chimpanzee has been raised in a human home and then abandoned; she then meets a group of fake chimps who subtly persuade her to return to the wild. The story was inspired by reading Roger Fouts' *Next of Kin*, about the sign language experiment with Washoe who, like other enculturated chimpanzees, fell somewhere between humans and chimps in terms of behaviour and identity. I felt that might resonate with chimpanzees in captivity. Given my knowledge of the chimpanzees' limited attention span for video, I thought the plot would ultimately not matter to them. The film would consist of exciting moments for the chimpanzees and a plot for the humans.

The process continued with the presentation of the finished 22-min film to the chimpanzees and the recording of their reaction. The project was presented as a two-channel installation, but it was difficult to watch both films at the same time. So I went a second time with an editor, Augie Robles, to create an experimental single-channel film. It was all an experimental project!

4.3.14 *Creatures*. Andro Montoya Riveros, 2008

By Andro Montoya Riveros

For more than 15 years I have been invoking creatures from my inner universe, most of the time I use spray cans, also brushes and rollers impregnated with latex paint or water-based enamel, and almost always, they remain on the walls of streets and neighbourhoods in Santiago and other places in Chile (Fig. 4.14).

I have been drawing since I was a child. My father trained me by making lines on paper that I had to transform into something meaningful. I felt satisfied when I managed to make something that seemed to be an animal or a living being.

I call them invocations because, unlike many fellow street painters who have their sketchbooks swollen with various designs, I hardly ever sketch the creatures. On the day I will paint the mural on the wall with my materials and stand in front of it not knowing what I will do. They appear at the moment, I do not plan them, probably because when I started out I was writing my thesis while working in a molecular neurobiology laboratory where everything was very planned, each experiment controlled, and even some with expected results. Therefore, the weekend when I painted was a time for improvisation in front of a wall, a time to let things flow, a visceral expression of sensations, generally expressed in the first stroke, a long one that defines the vertebral axis of many creatures. If there is tension and metal music in my headphones, a predatory-looking creature with a harder, straighter line, sharp edges, spikes, claws and fangs might appear. If I´m listening to reggae or music like it, maybe I'll start with a sinuous line of fluid curves and imagine following it up with long and delicate fins, transparent maybe, multicoloured for sure. And that's how the creatures appear, that's how they even surprise me. That moment of invocation in strokes, when the creature first emerges, is one of my favourite moments in the creative process.

I frequently have a direction I want to follow, an intention to look for a kind of marine animal, or a creature with ant legs and a scorpion tail, for example. Anyway, the plan may change along the way, that is allowed. I don´t let the plan define what guides the specific choice of a creature to invoke. Sometimes the process is more bodily than the vertebral tracing at the beginning of each invocation, I think it's more like imagining yourself as that creature. Once my great friend Defos (Guillermo Mardones) made me think about it, and I told him that if I were a creature, I would like to be like one of the ones I paint. And it

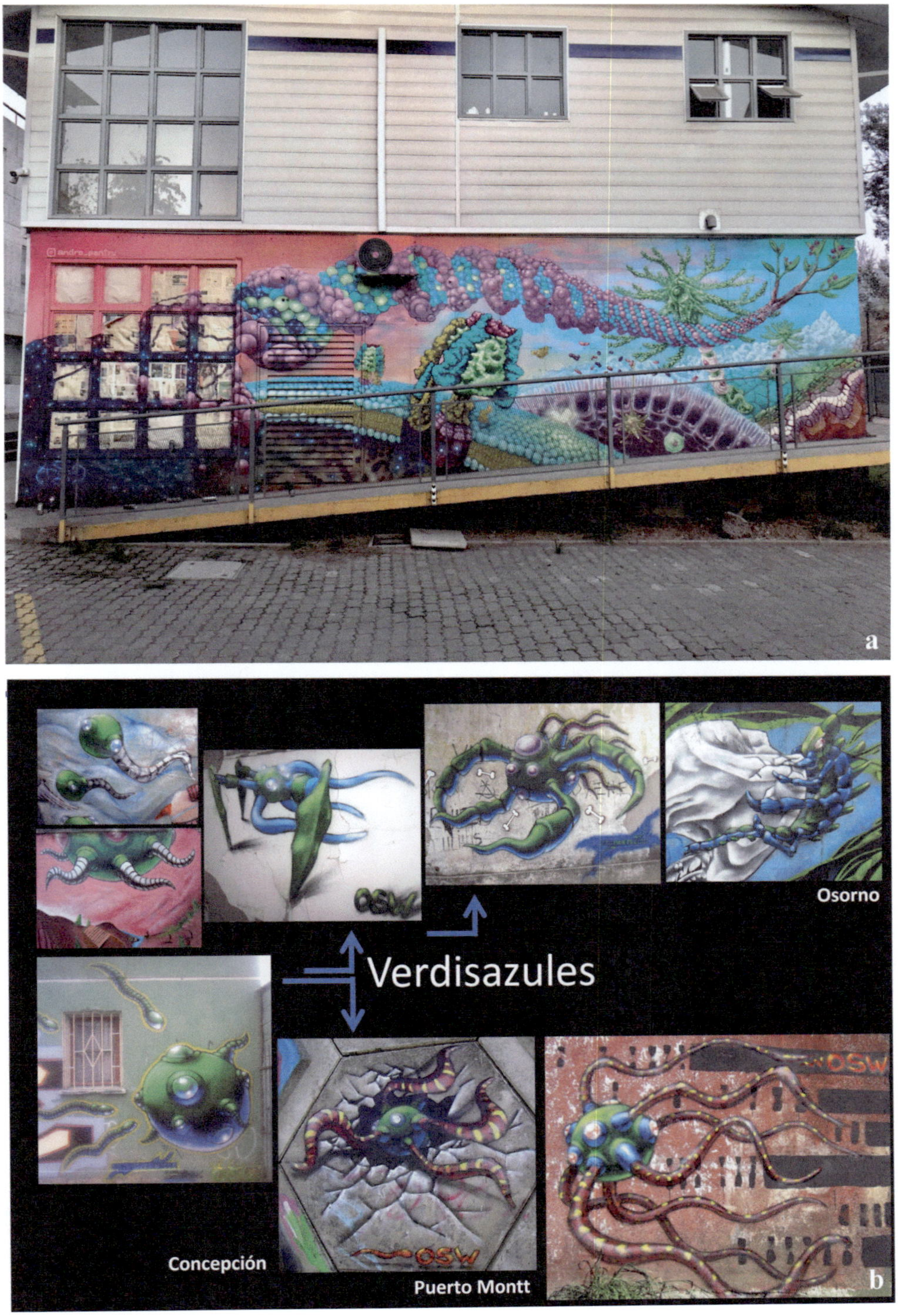

Fig. 4.14 **a** Andro Montoya Riveros. July 2021. *Sueño celular*. Facultad de Ciencias, University of Chile. Macul District, Santiago. Image courtesy of the artist. **b** Andro Montoya Riveros. 2008–2011. *Desarrollo de los Verdisazules*. Chile. Image courtesy of the artist. **c** Andro Montoya Riveros. November 2018. *Micro creaturas terrícolas*. Congreso de la Asociación Latinoamericana de Microbiología, Santiago. Image courtesy of the artist. **d** Andro Montoya Riveros. December 2021. *Monstruo celular*. Renca District, Santiago. Image courtesy of the artist. **e** Andro Montoya Riveros. September 2018. *Ciervo de cristal*. City of Coquimbo. Image courtesy of the artist

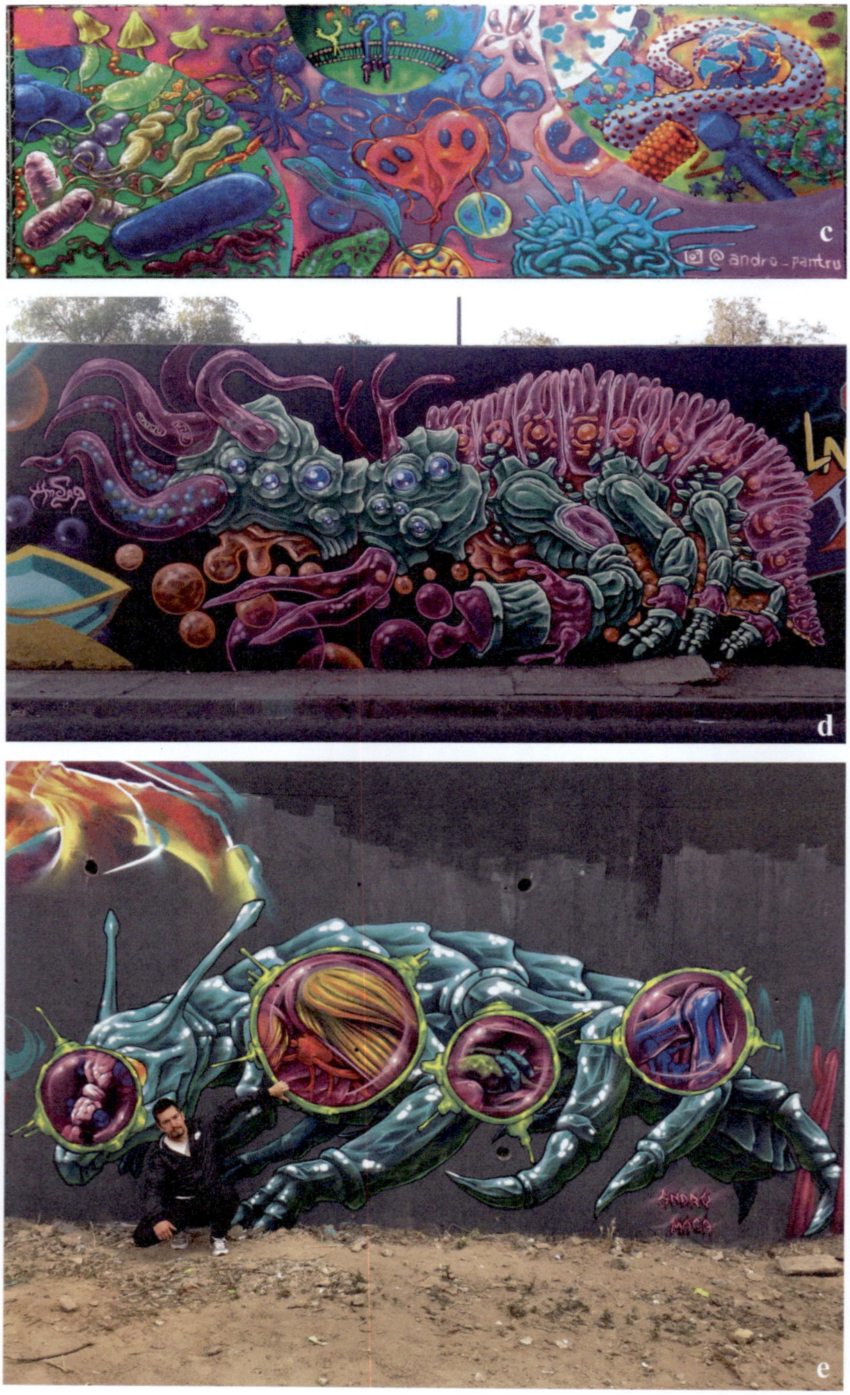

Fig. 4.14 (continued)

seems that I don't look for perfection in that new being of latex and aerosol, because it never stays, another one always appears. I find this diversity attractive, never repeating the same creature, always a new one, to increase the collection, so as not to stop the evolution.

However, over the years, some invocations seem to be very similar, at least in basic shapes and colours. Invoked in time and space sometimes very distant from each other, but I do not think they are the same creature. I apply my knowledge of biology to analyse them and discover the reason for their similarities: they are not the same creature, but the same species in different stages of their ontogeny, their biological development. In this way I can describe the invoked creature in more detail, feel it closer, because despite coming out of my inner universe, I don't know much about it. I feel like I know myself a little better after each new creature.

This has been a wonderful exploratory process, involving creatures, techniques, colour, neighbourhoods and streets, and lots of friendship and hip-hop music. My improvised creatures feel very comfortable in graffiti meetings or in rhyme-laden evenings with friends.

Having obtained my professional degree in Molecular Biotechnology Engineering, I decided to deepen my scientific training and entered a Ph.D. The demands of it, especially the first year, took me away from the walls for a considerable time. However, when I returned to painting, it seemed that I had also trained it. Now that I knew more about molecular biology, cell biology and neuroscience, something had changed in my inner world, and creatures began to emerge with more details in their constitution and structure. Transparent creatures began to appear, revealing their internal organs or cellular structures, definitely more complex organisms than before.

I was also interested in performing symbiote invocations. I call the mixture between humans and my creatures "symbiotes", a fusion that can be achieved with a lot of work and special body paints. To achieve this, I use the techniques of an art known as body paint.

After finishing my doctorate, I managed to get a job at a couple of universities to teach classes, and I began to teach others what I knew. Perhaps this natural and visceral flow from the inside to the outside allowed the need to communicate scientific knowledge, to use these public spaces to share science and health.

Now, my creatures are invoked with "scanned sectors", which makes it possible to see their internal structures, for example novel and bizarre organs, along with intricate skeletons and nervous systems. This is a new scale to analyse, study, classify, create and enjoy.

I have given some talks and lectures describing my art-science work in a variety of settings: in intellectual environments such as art and science symposia, in schools with young people and children, in institutions of scientific dissemination and even research laboratories. This has attracted the interest of scientists and popularizers eager to expose their specialties and knowledge.

Recently, together with my friend Defos, a fellow painter and physician, we have been promoting an initiative for the communication and enjoyment of biological and health sciences. We create scientific info-murals that allow us to educate with our art, to build a kind of bridge between academia and society, and between the laboratory and the street.

This improvised way of invoking the inner self and expressing it in the environment that surrounds us is what I believe has generated these visceral and necessary mixtures in me. A sort of portal between these universes. I like to imagine what it would be like if everyone searched for creatures in their inner universes. I would like to encourage more curious people to use art and science as their tools of exploration.

4.3.15 *Solar Displacement.* Špela Petrič, 2014

By Špela Petrič

We live in a never-sleeping, insatiable society that subdues its members to its technological apparatus. Transcontinental corporations, the world stock market and increasingly mobile careers may be examples of structures of production, which prominently override the chronobiologies of its human agents, but in reality few escape the subtle pressure of "living life to the fullest" or achieving social success. Where lies the border of this culturally imposed skewing of the human biological rhythm, beyond which the primordial Zoë revolts against its bios and begins to devour itself? Which social installations sustain the population on the edge of its circadian tolerance? The project *Solar Displacement* (Fig. 4.15) is an agora for the deliberation between the acceptance of a culturally suboptimal body and the potential of human agents to emancipate themselves from their biology using alcohol, pharmaceuticals, natural remedies, or phototherapy. A custom application developed for the mobile phone keeps track of the luminosity of the participant's surroundings and mirrors the light conditions in the rats' environment. The rats respond to the artificial light by displacing their rhythm of activity as well as with self-medication. The synchronisation of the participants with their bioindicators thus allows them to observe their physiological bodies as if they were dissociated, autonomous entities and are simultaneously urged to take responsibility for the well-being of the rats' as well as themselves.

In mammals, a metabolic clock in the suprachiasmatic nucleus (SCN) serves as the central pacemaker, synchronising peripheral oscillators in other organs while also remaining sensitive to the environment, most importantly light. A special layer of light-sensitive ganglion cells from the retina convey information to the SCN with variable effects. The body clock can be discerned by following various signs, including periods of activity and rest. Stacking multiple days on top of each other, a visually recognizable pattern of daily rhythm emerges, sometimes very regular, other times with a continuous drift, or perhaps lacking a pattern altogether. In constant darkness—in the absence of the zeitgebers (environmental cues) of sunrise/sunset—living beings follow their endogenous clock runs at its natural speed. When exposed to light, bodies undergo a period of stressful adjustment. Even while staying in one place the majority of the population experiences some discomfort, since social schedules misalign with individual sleep preferences. The most common disturbances of the circadian rhythm are jet lag, late working shifts, delayed or premature sleep phase syndrome, and the syndrome of prolonged rhythm of sleep–wake cycle.

Solar Displacement aims to synchronise the circadian rhythms of each of the participants to pairs of rats living in the gallery. Each pair is housed in a ventilated box with ample water and food available at all times. The route of synchronisation is through the luminosity inside the cage, which changes according to the intensity of the light to which the human participant is exposed. After several days of coordinated exposure, the circadian rhythms of rats and the human are attuned and the pair of rats begins to act as an autonomous avatar of the biological body of the participant. Since rats are nocturnal animals, their activity cycle is shifted 12 h with respect to human activity, but the periodicity of sleep and activity cycles is comparable. Depending on the lifestyle of the participant, the rats may begin to exhibit some of the common symptoms of circadian rhythm distress. These symptoms are observed by monitoring the rat activity over time using a thermal camera mounted in each cage, as well as measuring their propensity to drink an extract of valerian, a remedy for anxiety and sleeping disorders both in rats and humans.

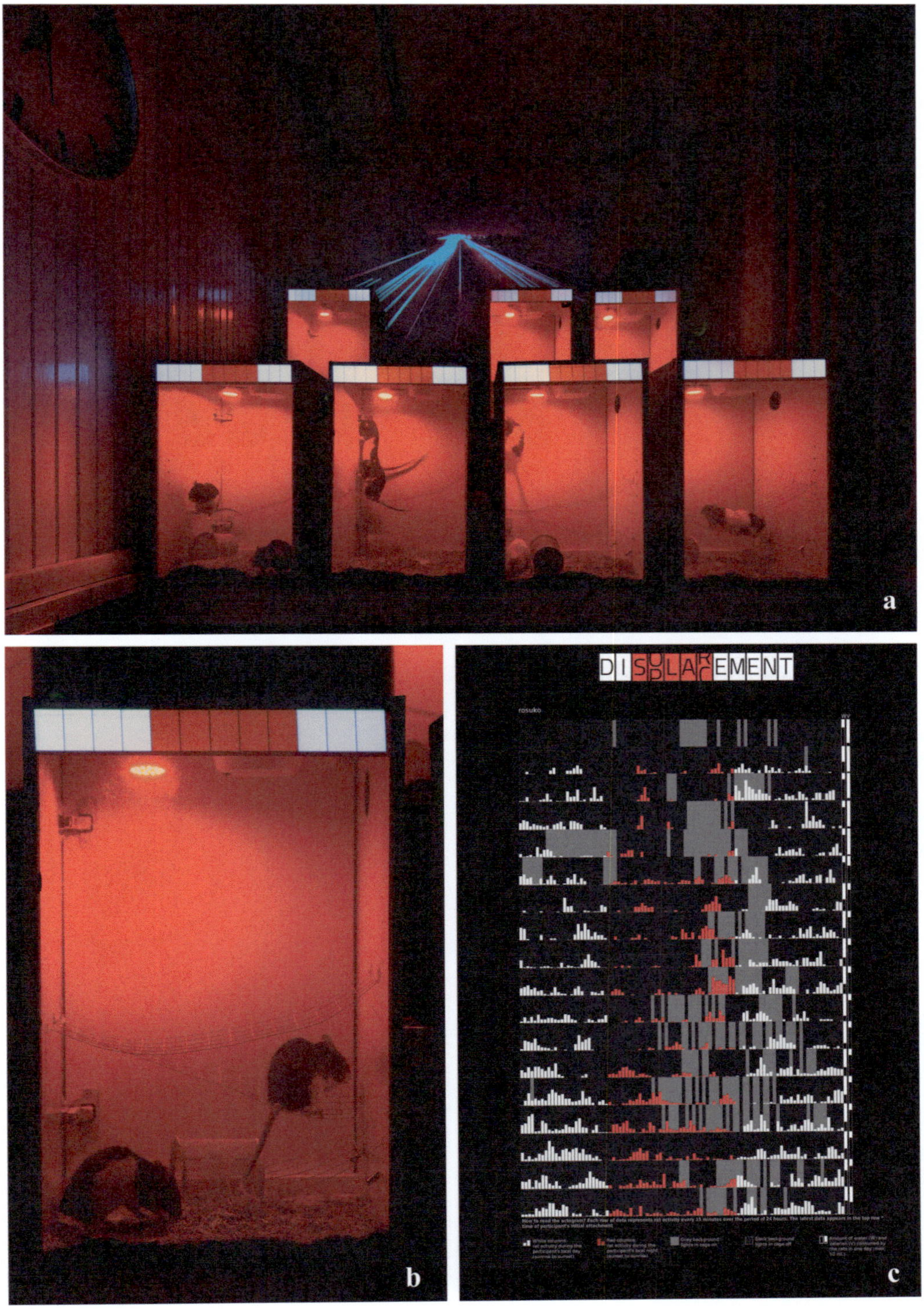

Fig. 4.15 Špela Petrič. 2014. *Solar Displacement*. Telematic Installation with living rats and responsive lights. Produced by Kersnikova Institute (Ljubljana, Slovenia). **a** General view. Photo by Damjan Švarc. Image courtesy of Kersnikova Institute. **b** Detail. Photo by Damjan Švarc. Image courtesy of Kersnikova Institute. **c** Rat's activity chronogram. Kersnikova Institute (Ljubljana, Slovenia). Image courtesy of the artist. **d** Rats Jerdt and Rosuko. Produced by Kersnikova Institute (Ljubljana, Slovenia). Image courtesy of the artist

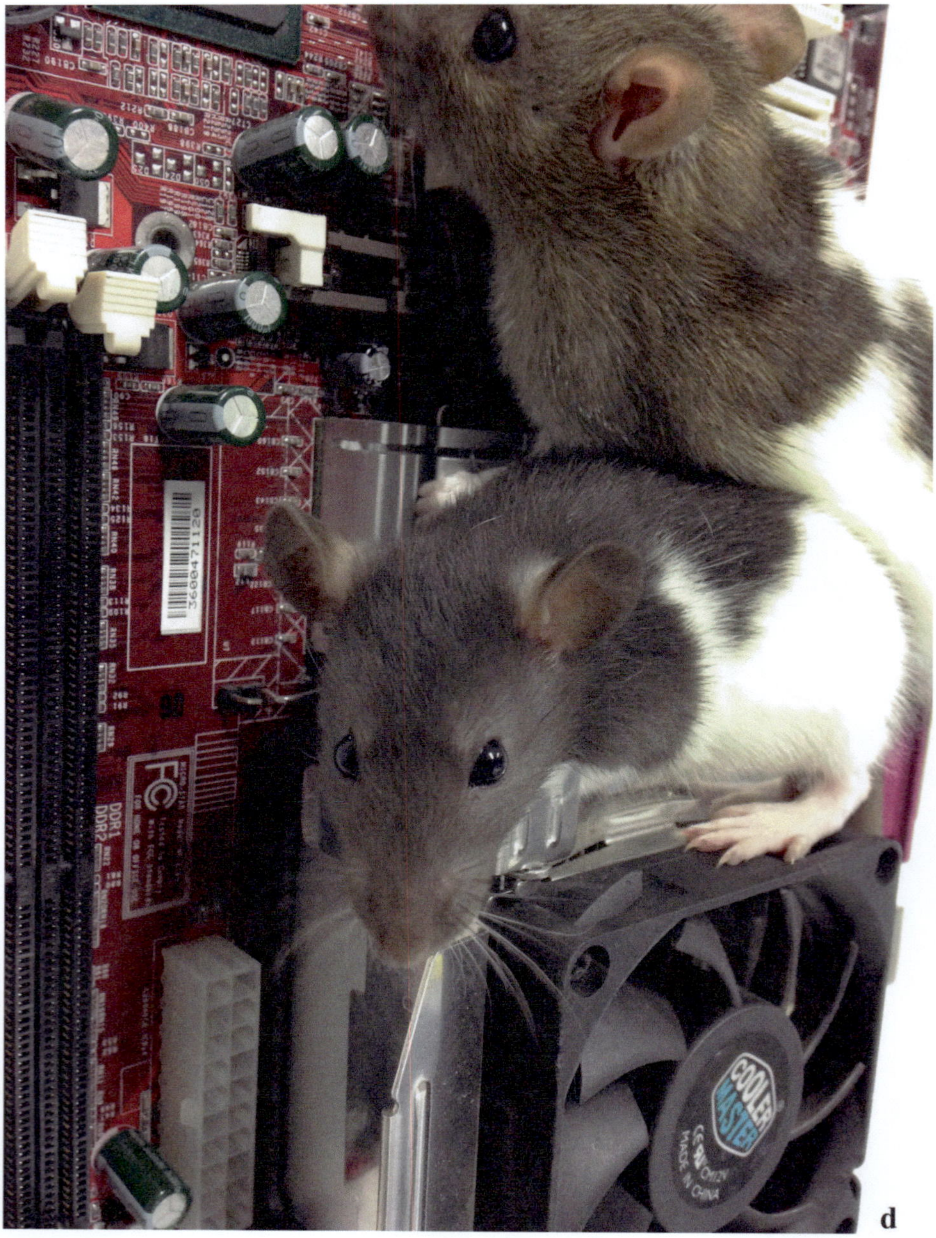

d

Fig. 4.15 (continued)

4.3.16 *Green, Grey or Dull Silver.* Perdita Phillips, 2007–2008

By Perdita Phillips

Green, Grey or Dull Silver: Art and the behavioural ecology of the Great Bowerbird, Chlamydera nuchalis was an Australia Council for the Arts funded residency at SymbioticA, an art/science research laboratory at The University of Western Australia. Fieldwork was undertaken between 2007 and 2008 at the Broome Bird Observatory. The Great Bowerbird is the largest member of the Ptilonorhynchidae family of passerines, that are most known for building elaborate, decorated structures or bowers, where males display to females looking for prospective mates. The primary function of the bower is as a display space for the male to exhibit its evolutionary fitness to the female.

The *Green, grey or dull silver* project (Fig. 4.16) researched aspects of the collection and display of objects by male Great Bowerbirds and bird vocalisations. It interrogated the processes of science at the same time as holding to an ethical relationship with the living world, influenced by multispecies thinking in the arts and humanities. There were three discrete but concurrent conceptual approaches in Green, grey or dull silver: the first was to approach it by adopting an aesthetic standpoint of 'interacting' or 'having a conversation' with bowerbirds. The second was to undertake the process of science (as faithfully as possible) with a view to adding a small but accurate contribution to the knowledge of science. The third perspective was to undertake the process of science and use this performative experience as a material for artworks.

The centrality of bowerbirds, bowers, display and collecting

The male Great Bowerbird builds an avenue of twigs and collects stones, bones, shells and human-made objects that he places in piles at each end of the avenue. The two heaps and the avenue in between are normally well hidden under low shrubs. The male performs for the female by calling noisily, posturing and dancing and by taking up the collected objects in its beak and waving them at the female. The female visits a number of bowers before choosing a mate. She watches the performance from the safety of the avenue and when she has made her selection, the two birds mate in or near the avenue.

Each species of bowerbirds has specific colours and shape criteria for the objects that the male collects. For the Great Bowerbird green, grey or dull silver objects are preferred. Often, he will take up green fragments in its beak (sometimes leaves but often glass), present his brilliant pink nape feathers to the female bobbing its head in a vigorous display. Different types of objects are grouped in specific areas of the bower and the male spends time each day arranging and rearranging his collection, renovating his bower, and even stealing from or wrecking the bowers of his rivals.

Wildness, entanglement and attunement

The relationship between people and the non-human was key to the project. There was a 'to-ing' and 'fro-ing' and a listening as well as a speaking in these relationships. The project used the framing of Spencer's (2004) conversational aesthetics—the idea that art is an informal co-constitutive conversation between the artist and their surroundings—to move away from representational strategies and to work with the way that non-humans actively inscribe themselves upon our world.

As wild birds in a wild environment, finding and interacting with bowers gave the artist an opportunity to come alongside the world of the Great Bowerbird in a way that respected their agency and attunement of the birds to their own habitats. It allowed a greater sense of 'equal energies' between bird and human—as birds took advantage of human things—and received or rejected gifts of exchange.

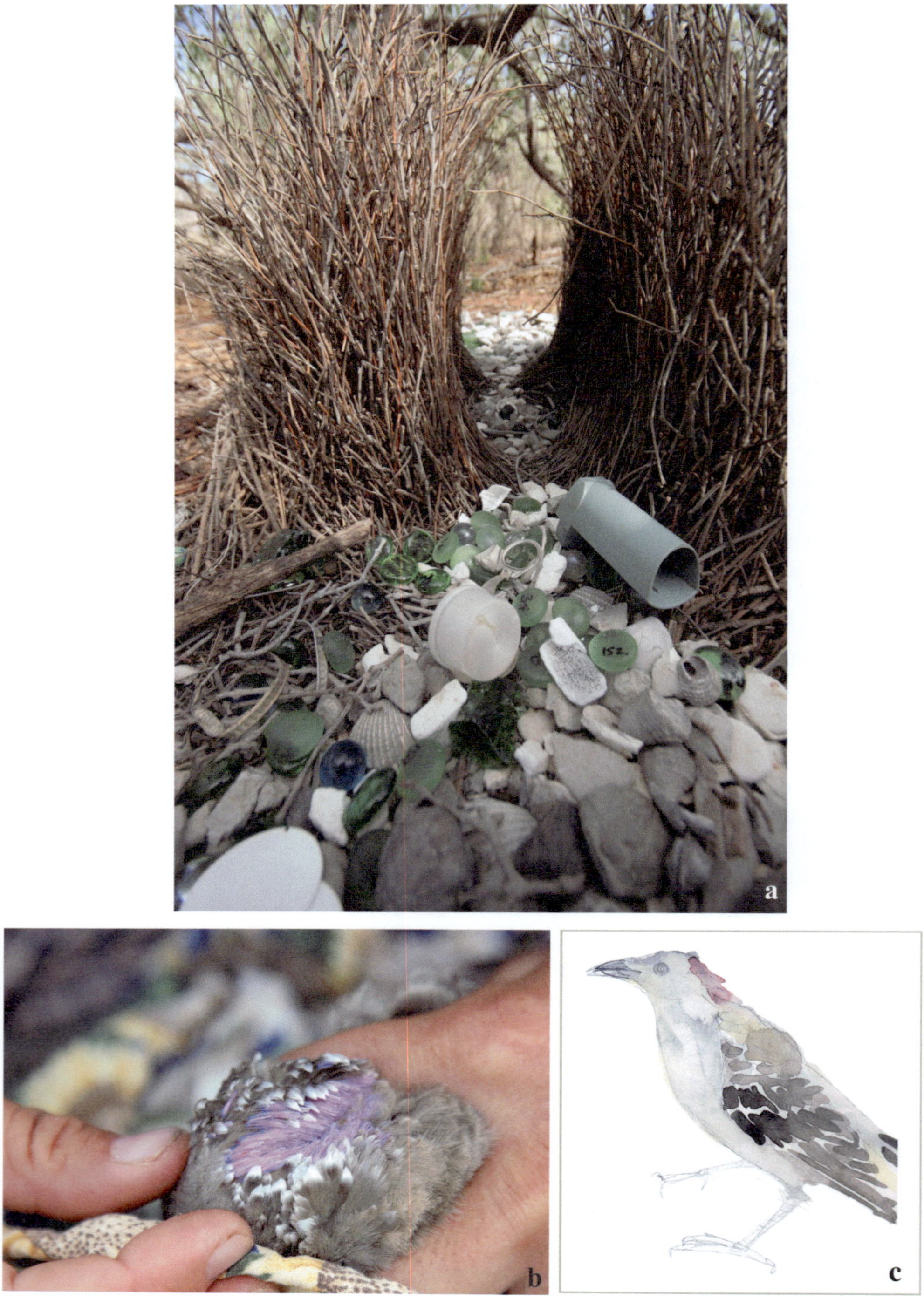

Fig. 4.16 Perdita Phillips. 2007–2008. *Green, Grey or Dull Silver.* **a** Documentation from Green, Grey or Dull Silver project: showing asthma inhaler and other materials gathered from a house nearby and green glass objects gifted as part of the interactions. Image courtesy of the artist. **b** Documentation from Green, Grey or Dull Silver project: bowerbirds were banded to be able to distinguish individuals. Here Class A bird bander Naoko holds the bowerbird to measure, weigh and band the bird; males (and occasional females) have an electric pink nuchal crest. Image courtesy of the artist. **c** Drawing from the *Green, Grey or Dull Silver* project. Image courtesy of the artist. **d** Documentation from *Green, Grey or Dull Silver* project: with gifted objects (green disks) moved from nearby to inside the bower avenue. Image courtesy of the artist. **e** Drawing from the Green, Grey or Dull Silver project. Image courtesy of the artist

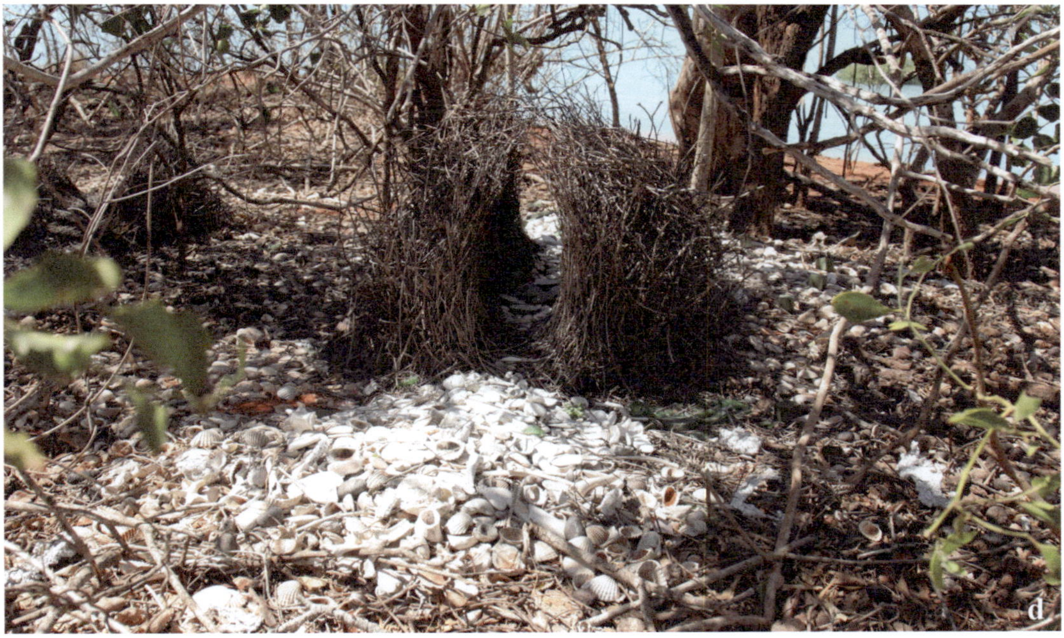

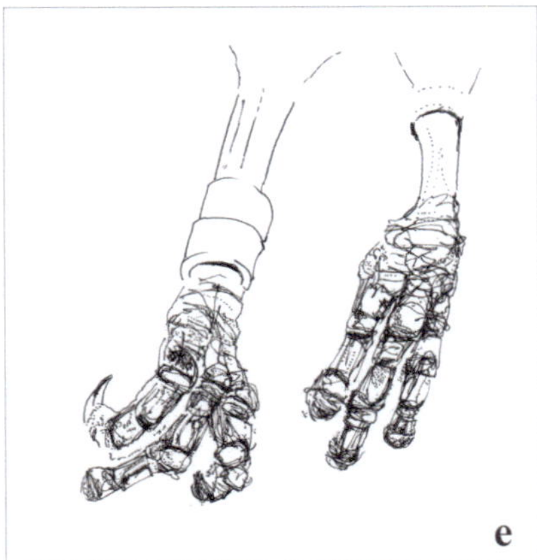

Fig. 4.16 (continued)

Issues and outcomes

Some of the issues encountered during the fieldwork include first finding and then visiting bowers (see Phillips 2012). The unpredictable behaviour of the weather in Broome led to the birds not being active or responsive to the first experiments of objects presented to the bowers—it was too early in the season. It took another field season to complete this part of the project. Dealing with wild animals meant that the initial objectives might have been overambitious. Attempts to influence mimicry (at a sound station) were not successful but the exchanges of objects did finally lead to results. Determining exactly the shape, colour and material the bowerbirds were most likely to interact with took some trial and error. The males freely availed themselves of human-made objects as long as they fit certain criteria of colour, size and roundness. Other human materials included light bulbs, plastic soldiers, marbles, aluminium foil, a spoon, golf balls and asthma puffer packs. But it was more difficult to trial shapes of objects consistently. Finally, changing from grey to green objects resulted in the gifts acted as 'conversations' with the male bowerbirds. The selection criteria of bowerbird choices are thought to be both genetically inherited and (perhaps) culturally learned and transmitted. Madden (2008) takes a definition of behaviours that support animal culture as being learned; learned socially; normative and collective. He argued that vocalisations, bower orientation, design, decoration use and display movements amongst the bowerbirds might fulfil these criteria, and thus constitute a 'culture'.

A concern was whether observation and measurement of bowers were going to have long-term effects on bowerbird behaviour. Fortunately, Great Bowerbirds are a gregarious species and the males returned to their original behaviour within minutes of humans departing from their bowers. The bowerbirds do not think of us as unique or special entities. As individual organisms they carry around an umwelt, a perceptual world within which things 'fit'.

To experiment with ecosystems is to have your effects sent outwards into the world. Returning to the field in 2008, green glass nuggets that had been placed at a central point, slowly began to be radiated out to different bowers in the landscape. It is likely that this pattern of spread and transmission over a wider area has continued over the years since. This is one of the challenges and benefits of working outside of laboratory-based sites with the unique set of spatial and temporal scales and complexities of ecology, of the combination of forces and processes and living and non-living things that make up ecosystems. Ecological studies of natural systems encounter many variables that cannot be easily controlled. Many environmental and eco-artists support a 'balanced' nature, favouring harmony over chaos and change but most ecologists (and some aggressively so) would support a changing and unbalanced nature, utilising concepts such as resilience (see Phillips 2010, 2015).

Bowerbird object classification was explored as part of *The System of Nature* exhibition (Lawrence Wilson Art Gallery, The University of Western Australia 2007) celebrating the 300th anniversary of Carolus Linnaeus. The sound installation, *The world has no shortage of things* (Phillips 2007), had both an ambient soundtrack through the entire gallery as well two opposing speakers comparing bowerbird calls with taxonomic descriptions of the birds. These two sound sources were combined with two shelves in a narrow corridor that contrasted collections of objects actually collected by bowerbirds with geometric shapes, the latter alluding to Linnaeus's failed attempt to extend his system of classification to the mineral kingdom. The artwork created an insistent 'press' upon the membranous border between human and nonhuman worlds.

References

Madden J R (2008) Do bowerbirds exhibit cultures? *Animal Cognition, 11*(1), 1–12. https://doi.org/10.1007/s10071-007-0092-5

Phillips P (2007) *The world has no shortage of things (the world of the Great Bowerbird)* Retrieved from https://www.perditaphillips.com/portfolio/the-world-has-no-shortage-of-things-the-world-of-the-great-bowerbird-684/

Phillips P (2010, 28 November) *Why unruly?* Retrieved from https://www.perditaphillips.com/portfolio/why-unruly/

Phillips P (2012) Fieldwork with bowerbirds. *Antennae: The Journal of Nature in Visual Culture, 20* (Spring), 98–105. Retrieved from http://www.antennae.org.uk/back-issues-2012/4583477699

Phillips P (2015) Artistic Practices and Ecoaesthetics in Post-sustainable Worlds. In Crouch C, Kaye N and Crouch J (eds.) *An introduction to sustainability and aesthetics: The arts and design for the environment* (pp. 55–68). Boca Raton, Florida: Brown Walker Press.

Spencer R (2004) *Hermes Gift: Walking, phenomenology and conversational aesthetics.* (Doctor of Creative Arts). Curtin University of Technology, Perth.

4.3.17 *Inhabiting a Map.* Michail Rybakov, 2018

By Michail Rybakov (Fig. 4.17)

In the year 2018 I was doing a two-years long postgraduate artistic research on the topic of 'Strategies of arrival'. It was fun, as it allowed me to go back to very basic questions—"what happens when I enter a room?" or "what happens when I enter a room and forget why I went there?" or even "what happens when I enter a room after a long, long trip, and feel weird for a few seconds until I fully arrive?". The research generated quite a few exciting artistic experiments in the field of psychogeography, not least because my mentor during the postgraduate research phase was the French sociologist and philosopher Bruno Latour. At the time he was preparing his exhibition on 'Critical Zones' at the ZKM Karlsruhe and was interested in how we could describe the processes of life as well as our very human relation to them. How to show all that is happening in the Critical Zone?

The Critical Zone is all about the processes of living, about living organisms and their entanglements. The difficult thing about dealing with the Critical Zone is that our usual mental models that describe the space we live in seem inadequate and insufficient: top-down maps may show geographical features but are frozen in time. Worse yet, they don't show any chemical, biological, mechanical processes of living, or any living organisms at all for that matter. They capture a specific point of view, not a point of life (*point de vue* versus *point de vie* as Bruno Latour would say). Sure, maps might sometimes show a single tree, but not the other things that make the tree alive—the water level, the rain, the chemical exchange between tree roots and bacterial colonies, the parasites and the birds that hunt them and so on. There is no future nor past in a map, there is no illness, no movement, no birth or death in a map.

It's not a problem of scale, as it does not change even if we become very local, if we zoom in closely on one specific point. For example, we can go far above ground, onto the near-earth orbit where the satellites are gliding through the darkness until they fall down and burn up during their fall. In that orbit we could stay for thirty years and observe Dubai between 1984 and 2016, observing everything that happened there. We would see an extreme example of terraforming and land development, fuelled by vanity and the political will to build funny looking shapes into the ocean. And even with this being the extreme example, the timeframe of thirty years is a large one. A casual observer drinking tea in orbit and looking down on earth in real time would not notice the change brought by our human processes of life. The distance is just too grand.

Imagine another scene—a flock of common wood pigeons landed in front of my house in Karlsruhe, February 2021. The polar vortex had broken in February, covering large parts of the US and Europe in ice and snow. The wood pigeons moved from the woods to cities to find a bit more warmth and food. I observed them in the snow—they moved as individuals, until they got spooked by an arriving bike driver. They moved at once and flew away as a flock, as a superorganism.

This slice of life may be human scale, but it still shows just one point of view. The text description above is needed to provide context, to turn it into a (very limited, admittedly) point of life.

It's difficult to think of 'points of life'. We have become accustomed to maps, to seeing our position from the outside, from above. But that position only shows the geographical proximity of things. It doesn't show things that are important for me, for my sustenance. Perhaps earlier, a few hundred years ago, a small map of a village would be enough to depict most of the things a villager would depend upon in order to survive—a water well, fields, a river, a mill, a cowshed, and the whole community of the village that could survive for a long time without direct interaction with the world outside the village. This image of such a village, however, a village that lives in itself, has not been the lived experience for a long time. Everything and everyone is made up of incredibly complicated interconnections. And where before we could ask the villager

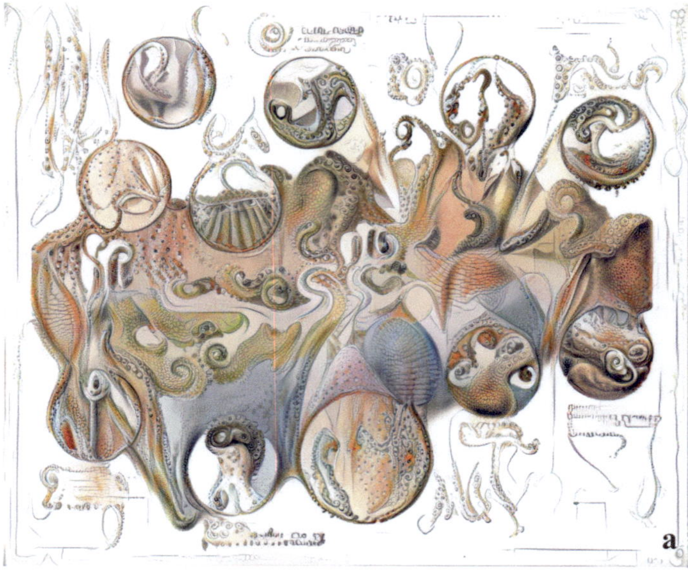

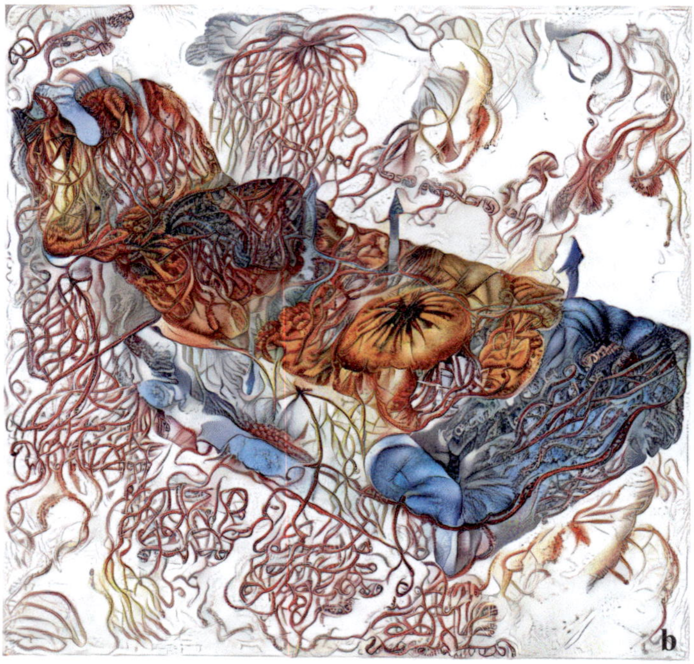

Fig. 4.17 **a** Michail Rybakov. 2018. *Gamochonia, Ernst Haeckel versus Climate Change in California, Department of Water Resources*. AI-generated cross-section diagrams populated by the animals that may live there. From the project "Inhabiting a map" for the Critical Zones research seminar with Bruno Latour. Image courtesy of the artist. **b** Michail Rybakov. 2018. *Siphonophorae, Ernst Haeckel versus Urban drainage, U.S. Geological Survey*. AI-generated cross-section diagrams populated by the animals that may live there. From the project "Inhabiting a map" for the Critical Zones research seminar with Bruno Latour. Image courtesy of the artist. **c** Michail Rybakov. 2018. *Siphonophorae, Ernst Haeckel versus Urban drainage, U.S. Geological Survey*.AI-generated cross-section diagrams populated by the animals that may live there. From the project "Inhabiting a map" for the Critical Zones research seminar with Bruno Latour. Image courtesy of the artist. **d** Michail Rybakov. 2018. *Nudibranchia, Ernst Haeckel versus Principles of relative dating*. AI-generated cross-section diagrams populated by the animals that may live there. From the project "Inhabiting a map" for the Critical Zones research seminar with Bruno Latour. Image courtesy of the artist

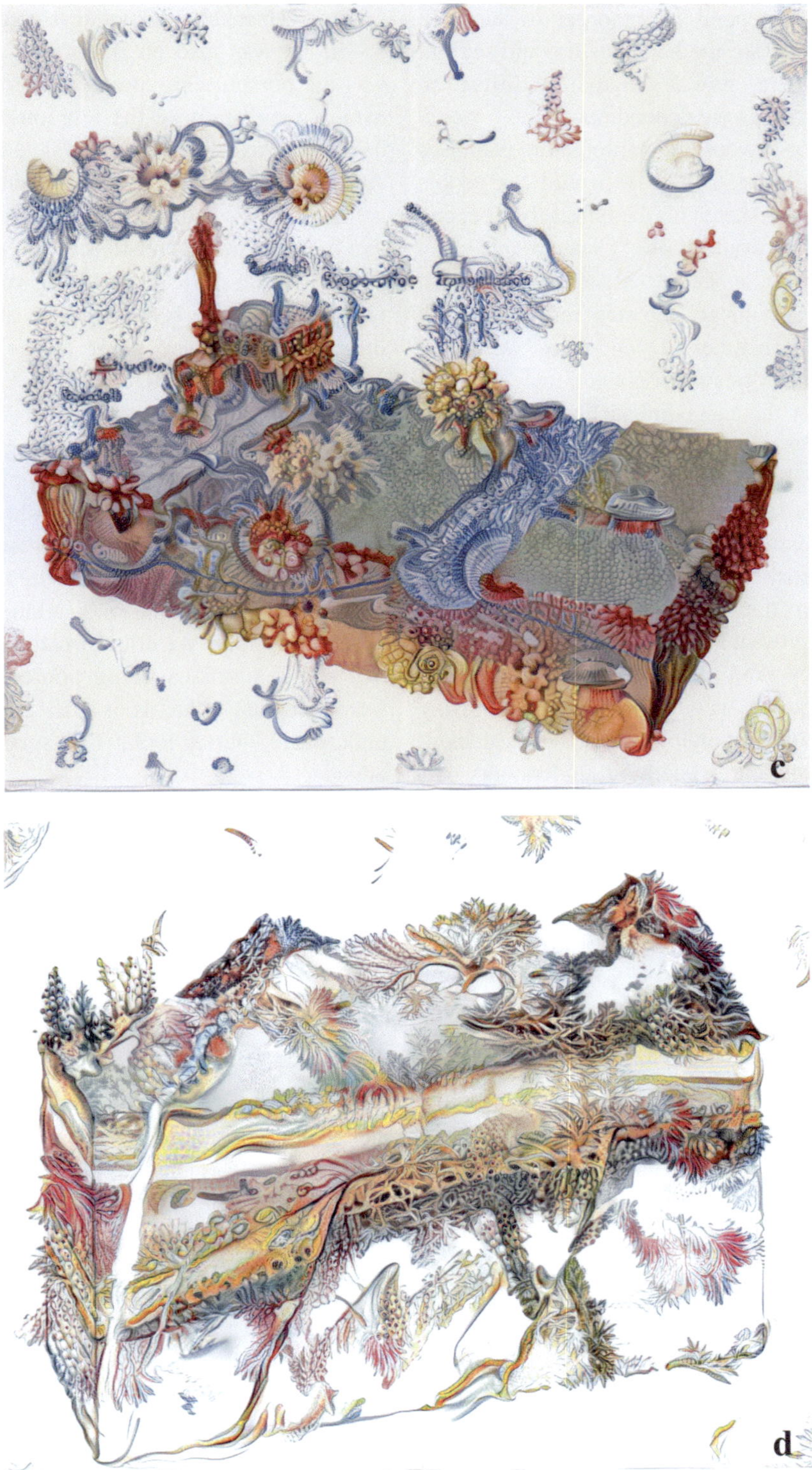

Fig. 4.17 (continued)

'what do you depend on in order to survive', and he would make the hand gesture and say 'all that you see here', now it's incredibly difficult to count all the things we depend upon.

Bruno Latour would do participatory exercises in many places he visited. He would ask people to write down or sketch all the things they depend on. The person would write "Clean water; clean air; tasty bread; avocados; a shelter…" and so on. Then he would go one step further and ask what does "clean water" depend upon? Well, that would be a water treatment plant, people working there, industry that can build such a plant, a union that represents the needs of the water treatment workers, finally, all their needs and dependencies too. So, a whole network of actors would develop from a single question. We could call it an Actor-Network Theory.

Trying to translate such networks into a graphical representation, is, of course, an exercise in futility. And this, as I see it, is where art comes in. Art as a way of thinking, art as a way to allow the unexplainable remain unexplained, while still making something understood. Art can exist outside of inductive/deductive reasoning and still 'make sense'. Call it 'artistic reasoning' if you must.

My way of artistic reasoning was to try and play with the complexity of maps. Maps are expected to be simple, they are expected to show exactly the information you need but nothing else. So, I decided to add more information to the maps and illustrations of the world, more than is necessary for humans to read them. In order to achieve that I turned to Ernst Haeckel and his beautiful illustrations of living organisms. I would combine these illustrations—or some quality of these illustrations with the structure of earth cross-section diagrams often used in geological sciences. Both the illustrations of organisms and the cross-section diagrams aim to explain a quality, a character of an animal, or of a place in which the animal resides. I used a Neural Style Transfer algorithm to combine the qualities of both these images into one, combining both the animal and the ecology that surrounds that animal.

Ernst Haeckel was not just a talented illustrator; he was also an important biologist. He was the person who coined the term 'ecology', meaning the study of the relationships between living organisms and their physical environment. And now, his illustrations were being used to generate an extended view—a view of ecology that includes the organism.

For Neural Style Transfer to work you need two images—one content image and one style image. From the content image, large features are preserved. Onto these large features the smaller features from the style image are imposed. For example, if we take an image of a chair as a content image and a photograph of a cactus as a style image, it would go from the largest features towards the smallest, first making the chair greenish, adding the cactus spines over time, then, in the end, adding the spikes to it. Or perhaps we could imagine the process like sugar sprinkles being poured onto a still wet chocolate cake. It is still chocolate cake underneath, but it looks like sprinkles on the outside!

The Neural Style Transfer works in multiple runs—you can imagine it as a florist cutting down and rearranging a way too exuberant bouquet—after each iteration looking again at it from the side and seeing if it is getting closer to a mental image of a perfect bouquet, then deciding what to do next. Depending on the strength of the cuts and changes, you may need fewer steps, but you may also 'overshoot': making too many changes or missing out on small details that would help the overall appearance. Neural Style Transfer can be tuned to an array of settings, like the aggressiveness of changes, the strength of the content image and the strength of the style and the number of runs. There is always some balancing involved, and no resulting image is ever 'perfect'.

To test things, I would run the Neural Style Transfer with different settings at different resolutions. Once I had the results I liked, I ran all the 'style' images on all the 'content' images. There were pairs that worked really well, and some others that didn't work at all. What you see was selected for expressivity and cohesion.

While I created many more works since, some dealing with similar questions of space, and body, and the not really very firmly-defined boundaries between them, this work remains one of my favourites. Here it all came together: the need for a new way of visual thinking; the very acutely felt choreographic fluidity of living things juxtaposed with the dead, non-living maps; the talented illustrator/biologist defining the term 'ecology'; the (then) state-of-the-art algorithm; the inspiring input of Bruno Latour and my peers. What came out of it feels like more than the sum of its parts.

It is, for me, a very lively example of a role art can play in science, in adding a new way of thinking to our arsenal. When you're stuck in complexity, just use artistic reasoning.

4.3.18 *Bee's*. Susana Soares, 2007–2009

By Susana Soares

Scientific research has demonstrated that bees have an extraordinarily acute sense of smell and have the potential to be used as an effective diagnostic of disease. What if we could use insects' extraordinary sense of smell to prevent and diagnose illness?

The project *Bee's* (Fig. 4.18) harnesses the olfactory abilities of honeybees to perform a health screening test on people's breath odour.

Bees can be trained within minutes using Pavlov's reflex to target a wide range of odorants associated with certain diseases [1]. The training resorts to their natural food foraging behaviour and consists in baffling the bees with a specific odour and feeding them with a solution of water and sugar, associating the odour to a food reward.

A conditioned bee or *sniffer bee* can detect, differentiate, and locate a scent at least down to parts per trillion level [2] which makes them a highly sensitive and accurate bio-sensor.

The glass objects depicted are a series of diagnostic tools that use trained sniffer bees to detect specific odours connected to marker compounds for diseases in human breath. The objects are designed with two enclosures: a smaller chamber that serves as the diagnosis space and a bigger chamber where previously trained bees are kept for the short period of time necessary for them to detect general health. The test consists of a user exhaling into the smaller chamber; the bees will rush in only if they detect the odour that they were trained to target. After the screening test, sniffer bees are released back to the beehive since the insect's welfare is an important aspect to safeguard.

But what exactly can bees detect? Studies have demonstrated that bees have the potential to detect tuberculosis, skin and lung cancer and diabetes [3]. The *Bee's* project is part of a speculative approach to design that intends to explore the role that design, and designers can play within the technological redesign of nature towards a sustainable future.

At the beginning of the twenty-first century, Paul Crutzen and Eugene Stoermer made a case for a new geological period in which humans and technological influence would dominate the earth. Since 2008, the 'Anthropocene' epoch has been widely accepted by the scientific community [4]. The impact that humans have had on Earth is undeniable and it is believed that as a result, humans have directly altered at least 70% of our planet's ecosystems.

Bee's explores potential symbiotic relationships with living organisms and how we might cohabit with natural biological systems in the future and use their potential to increase our perceptive and technological abilities. We have always co-existed with these systems, but their potential was unknown. This project is based on ongoing research (5) that provided the knowledge to enable new interactions. The project aims to develop collaborative relations between scientific and technological research, beekeepers, and design, among others, translating the outcome into systems and objects that people can understand and use, engendering significant

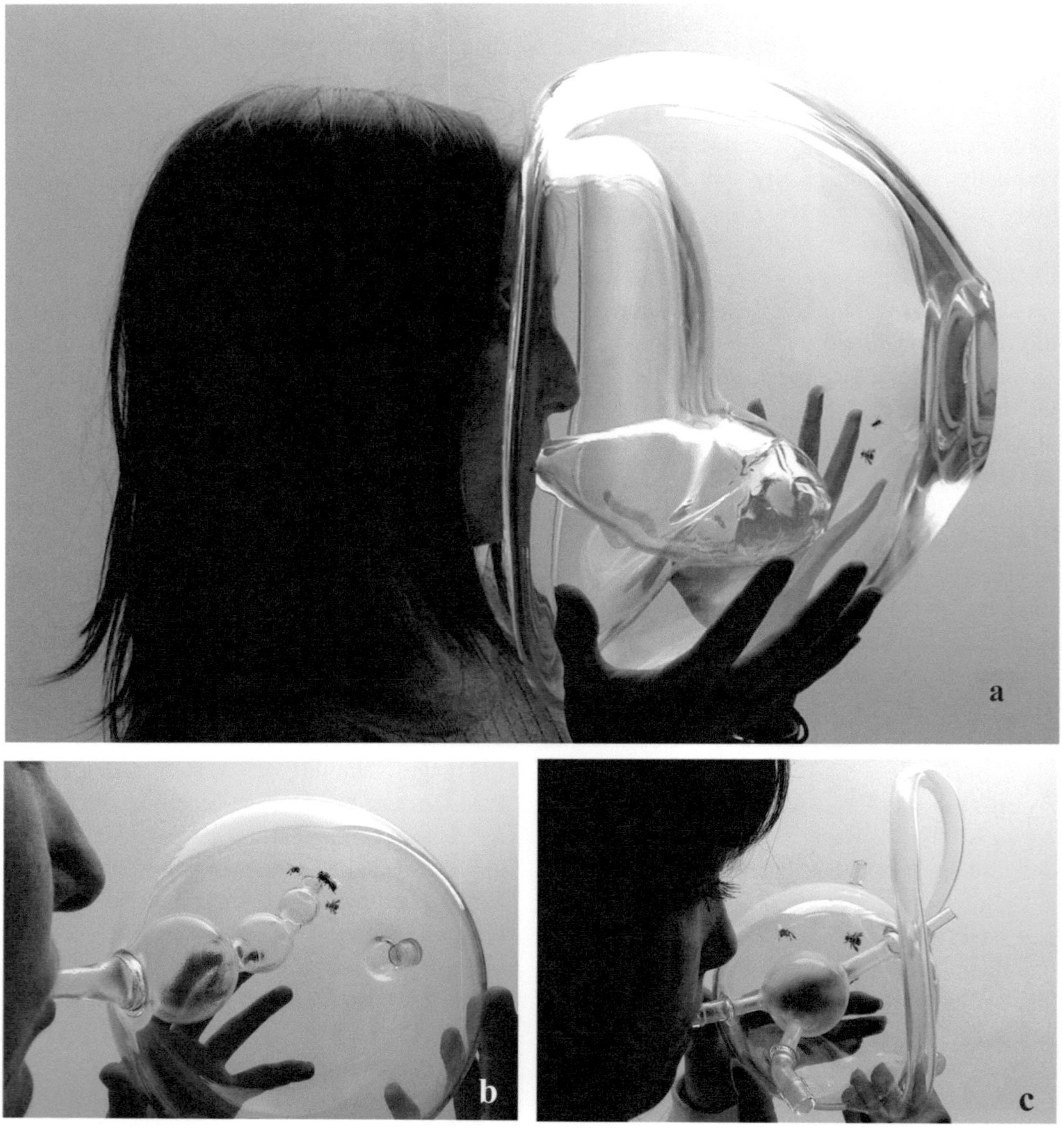

Fig. 4.18 Susana Soares. 2007–2009. *BEE'S*. **a** Face object: scientific research demonstrated that trained bees can detect certain diseases in a person's breath. The big chamber works as a bee container and the small chamber as the diagnostic space. The insects, previously trained to target a specific odour, react if sensing a familiar smell by flying towards the small chamber. Face object (36 * 25 cm) prototype 2007; Blown handmade glass; Crisform, Portugal. Image courtesy of Susana Soares. **b** Diagnostic tool 2: bees can be easily trained using Pavlov's reflex to target a wide range of natural and man-made chemicals odours including the biomarkers associated with certain diseases. Diagnostic tool 2 (26 * 15 cm) prototype 2007; borosilicate; Vilabo, Portugal. Image courtesy of Susana Soares. **c** Precise object: scientific research demonstrated that trained bees can detect certain diseases in a person's breath. Precise object (22 * 12 cm) prototype 2007; borosilicate; Vilabo, Portugal. Courtesy of Susana Soares. Image courtesy of Susana Soares. **d** Bee training object—bees can be easily trained using Pavlov's reflex to target a wide range of natural and man-made chemicals odours including the biomarkers associated with certain diseases. Bee training (20 * 7 * 9 cm) prototype 2009; clear acrylic & hips. Image courtesy of Susana Soares. **e** Precise object: at the beehive—checking the bees behaviour in the diagnostic Tools. Precise object (22 * 12 cm) prototype 2007; borosilicate; Vilabo, Portugal. Image courtesy of Susana Soares. Susana Soares. 2007–2009. *BEE'S*. Bee training object—bees can be easily trained using Pavlov's reflex to target a wide range of natural and man-made chemicals odours including the biomarkers associated with certain diseases. Bee training (20 * 7 * 9 cm) prototype 2009; clear acrylic & hips. Image courtesy of Susana Soares

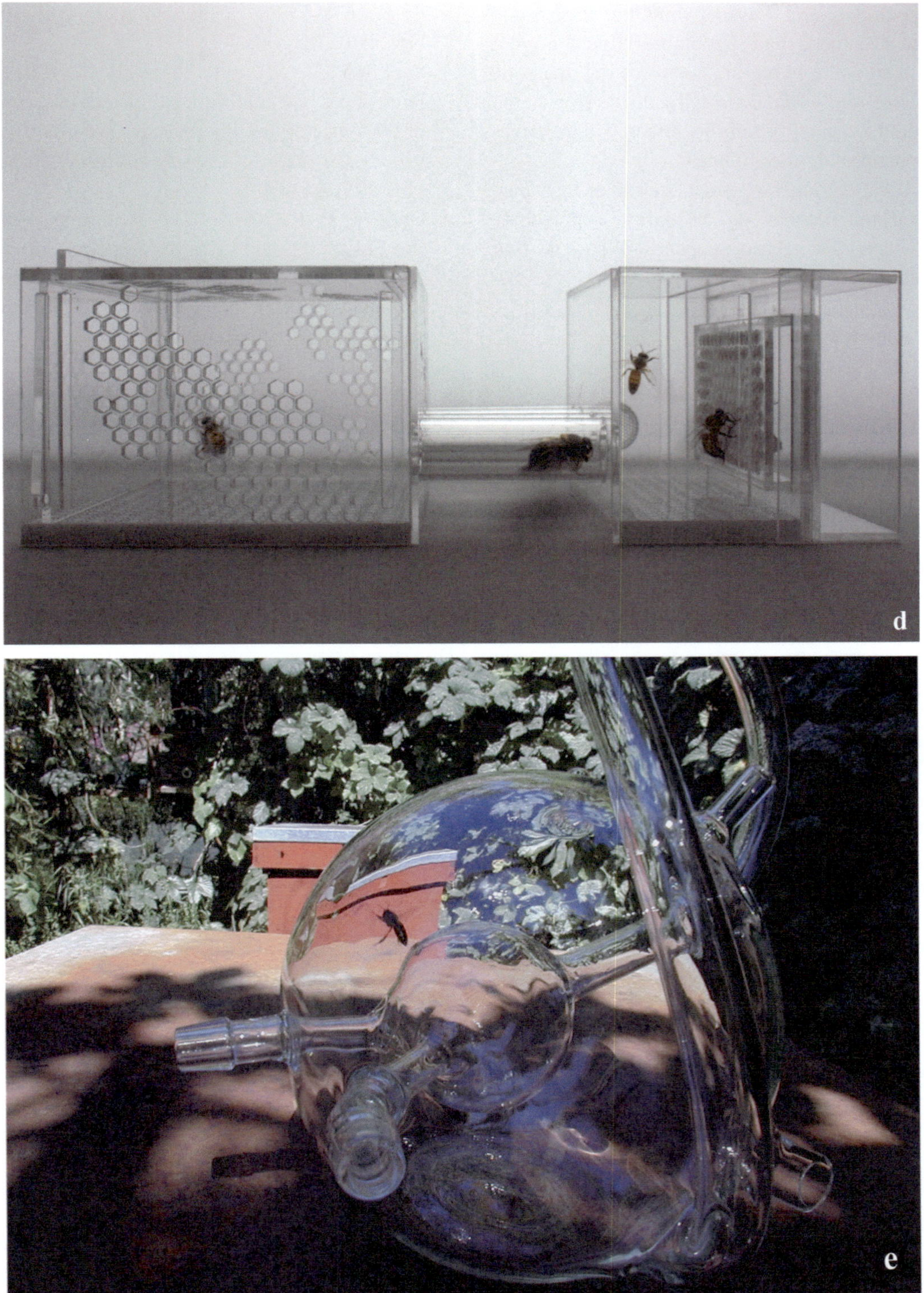

Fig. 4.18 (continued)

adjustments and ways of thinking. It also aims to spark the debate about the ethical implications of using living organisms as raw materials, '*engineered*' into certain functions for human benefit or using scientific research that harness the natural abilities of bees.

What should we trust more: a man-made electronic device or a highly sophisticated bio-sensor to detect cancer? What is more sustainable?

Acknowledgements

Calouste Gulbenkian Foundation

Prof. Anthony Dunne and Fiona Raby (Royal College of Art, Design Interactions Department)

Dr. Mathilde Briens (Inscentinel—UK)

Monica Santos (Filmmaker)
David Perkins (Roots & Shoots—London Beekeeper Association)

Des. Sónia Durães and Glass Master Mateus (Crisform)

Sr.João Gomes (Vilabo)

Credits: Susana Soares.

References

1. Rodacy P J, Bender S, Bromenshenk J, Henderson C and Bender G (2002) Training and deployment of honeybees to detect explosives and other agents of harm. *AeroSense*, International Society for Optics and Photonics, 474–481. https://doi.org/10.1117/12.479119
2. Bromenshenk J J, Henderson C B, Seccomb R A, Welch P M, Debnam S E and Firth D R (2015) Bees as biosensors: chemosensory ability, honey bee monitoring systems, and emerging sensor technologies derived from the pollinator syndrome. *Biosensors*, 5, 678–711. https://doi.org/10.3390/bios5040678
3. Suckling D M and Sagar R L (2011) Honeybees Apismellifera can detect the scent of Mycobacterium tuberculosis. *Tuberculosis*, 91, 327–328. https://doi.org/10.1016/j.tube.2011.04.008
4. Crutzen P J (2006) "The Anthropocene". In Ehlers E and Krafft T (eds.) *Earth System Science in the Anthropocene*. Springer, Berlín, Heidelberg. https://doi.org/10.1007/3-540-26590-2_3

Some of the laboratories around the world that have done research on crawler bees: Rothamsted Research, UK and CNRS, Paris, Inscentinel Ltd., Los Alamos and DARPA.

4.3.19 *Crocheted Hyperbolic Planes.* Daina Taimina, 2009–2014

By Daina Taimina

All these fibre sculptures (Fig. 4.19) are crocheted to form surfaces in mathematics known as hyperbolic planes representing two-dimensional surfaces with constant negative curvature, more properly described in my book Crocheting Adventures with the Hyperbolic Planes [1] or in Experiencing Geometry [2]. Hyperbolic planes can be also thought as opposite to a sphere which has constant positive curvature. Sphere is finite and closes on itself while its opposite expands exponentially. Spheres can be with various radii, so do hyperbolic planes. For The Land and the Sea, I chose to crochet a hyperbolic plane with a fairly large radius which starts to curl more when it gets to the blue/white part, representing a sea. This piece represents our concern for the Baltic Sea which is one of the most polluted marine ecosystems in the world.

Global Warm(n)ing is my largest crochet work and it was created in 2010 as a response to warnings about global warming which by now has already affected our planet. White Cloud and Crochet Continues are two of my collaborations

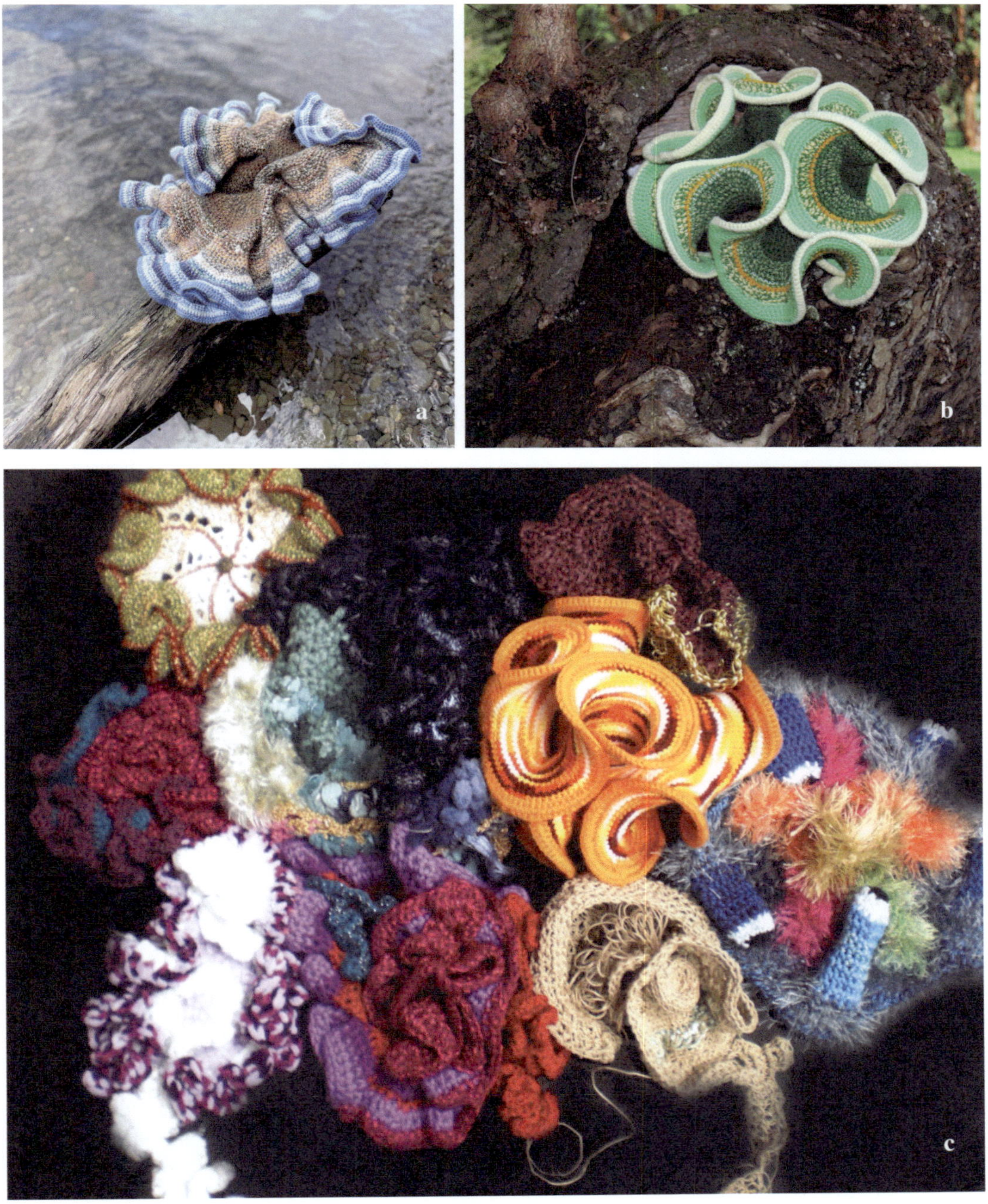

Fig. 4.19 **a** Daina Taimina. 2009. *The Land and the Sea*. Crocheted wool and acrylic yarn. Image courtesy of the artist. **b** Daina Taimina. 2010–2011. *Growing naturally*. Crocheted wool and acrylic yarn. Image courtesy of the artist. **c** Daina Taimina. 2011. *Crochet continues*. Crocheted wool and acrylic yarn. Image courtesy of the artist. Daina Taimina. **d** Daina Taimina. 2014. *White Cloud*. Crocheted wool and acrylic yarn. Image courtesy of the artist. **e** Daina Taimina. 2008–2010. *Global Warm(n)ing*. Crocheted wool and acrylic yarn. Image courtesy of the artist

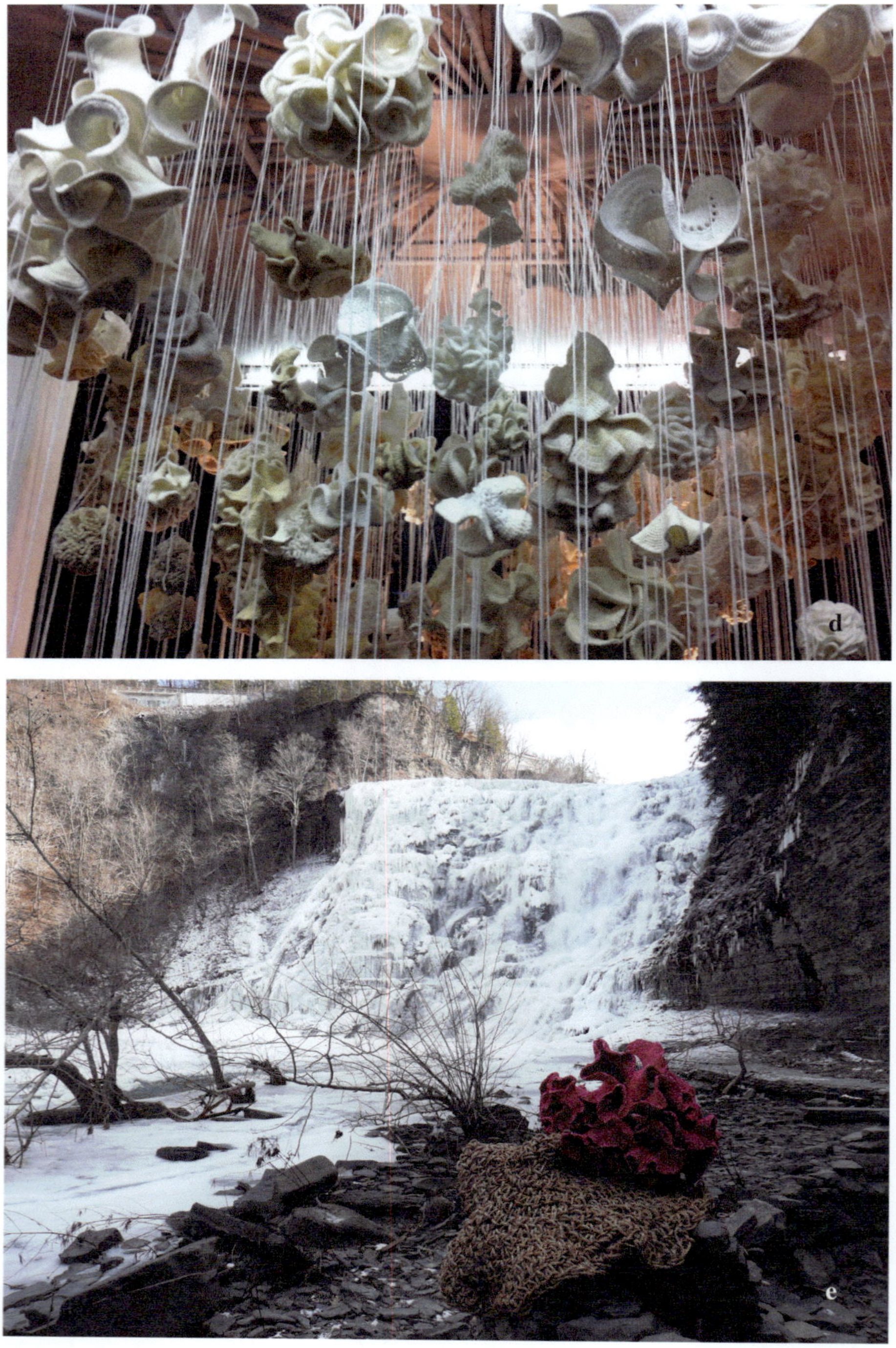

Fig. 4.19 (continued)

with many other crocheters. White Cloud was first exhibited in Riga, Latvia, in 2014, then travelled to Finland in 2016. Crochet Continues was first started in Sicily with one of my crochet pieces in the middle, then continued by passing it on all over Italy.

References

1. Taimina D (2018) *Crocheting Adventures with the Hyperbolic Planes*, CRC Press, 200 pp.
2. W. Henderson D and Taimiņa D (2020) *Experiencing Geometry: Euclidean and Non-Euclidean with History*. Books by Independent Authors, none: 358pp. https://doi.org/10.3792/euclid/9781429799850. https://projecteuclid.org/ebooks/books-by-independent-authors/Experiencing-Geometry/toc/10.3792/euclid/9781429799850. Free to download.

4.3.20 *Living Architectures*, Manuel Théry. 2008, 2013

By Manuel Théry

The work of Manuel Théry takes us on a voyage through the cell's universe. It combines micro-engineering, cell culture, fluorescence microscopy, image processing, video-projection and video-mapping (Fig. 4.20). The final video, projected on the facade of a building, shows cells whose internal architecture is adapted to that of the building. It is obtained by reproducing the building in miniature (one million times smaller), at the size of the cells, with micro-lithography, by placing living cells on it and by recording for several days the motion of inner cables which constitute the architecture of the cells. In the displayed movie on the building, the cell's internal architecture appears to sense and respond to the architecture of the building.

The aim of this project is to bring together different spatial scales, first by reducing a building to the size of the cells and then by enlarging the images of the microscopic cells to the size of the building. This highlights the similarities and differences between the huge architectures of buildings and the internal architectures of our cells. Viewers can see and compare the structures we make and those we are made of. They can appreciate the incredible differences between the geometries of living structures and those of inanimate buildings, between the natural rules of assembly within tiny organisms and those we have developed to build huge objects.

The building is miniaturised at the size of a cell (i.e. 100 μm = 1/10th of a millimetre), to place a human cell on this building and to video-record it on the microscope for several days. The cells had therefore to adapt their own shape to the building structure! Some cells move quickly and sneak around the windows, others, to the contrary, anchor themselves onto them and spread over the entire building. Cells were recorded for several days under a microscope, and the images were edited and video-projected back onto the original building. The cell thus gives the visual impression of literally adapting its shape and architecture to that of the building. All the details of the building are reproduced in the miniature replica. Therefore, the cell can feel and adapt to all these details. The internal architecture of the cell, its skeleton, consists of dozens of cables. All these cables get attached to every detail of the building (windows, balconies, stonework). The spectators thus see a giant cell stretching the cables that form its skeleton over the elements that constitute the building.

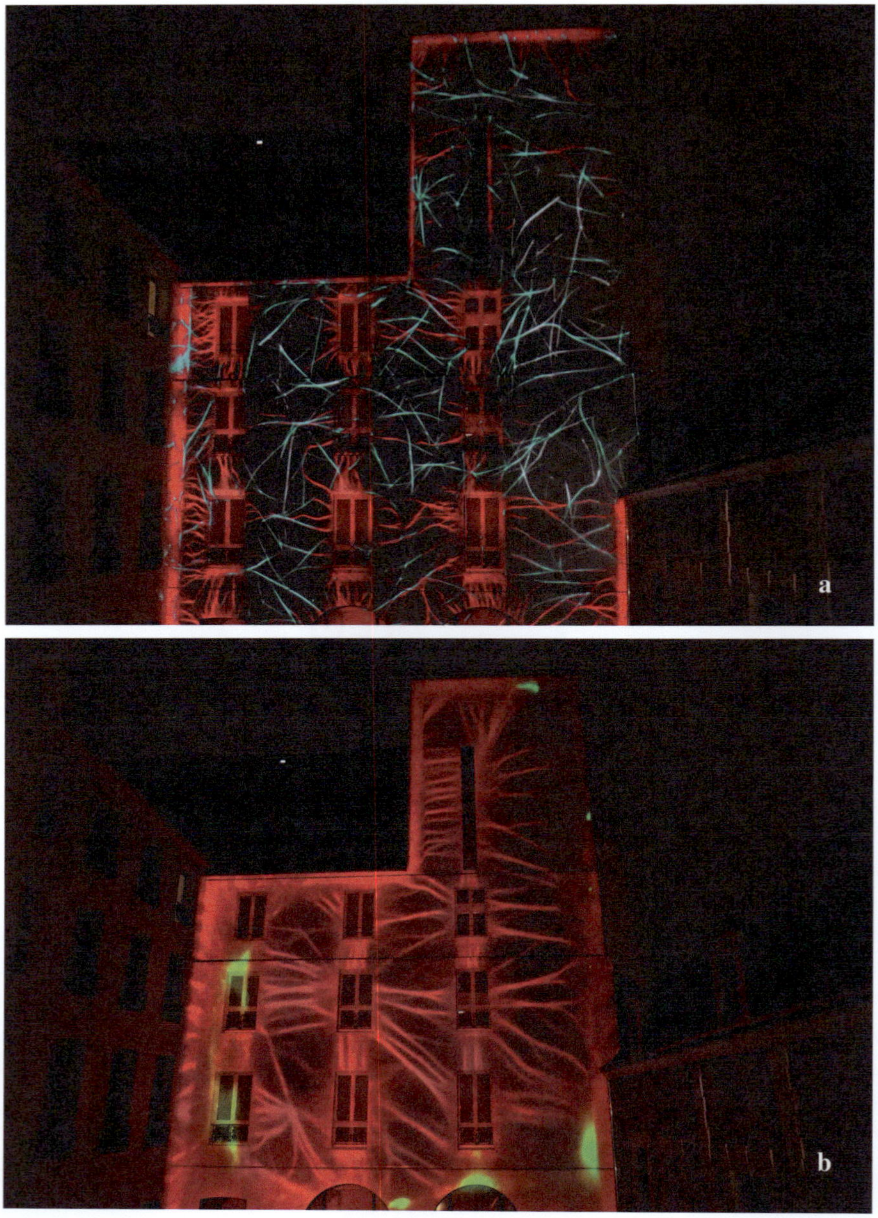

Fig. 4.20 Manuel Théry. 2008, 2013. *Living Architectures.* **a** Picture taken in 2018 of the building of the Centre de Recherche Interdisciplinaire, in Paris, during the Nuit Blanche. It shows the projection on the original building of an image taken on a microscope of the self-organisation of actin filaments (red) and microtubules (blue) on a miniaturised building at the micron scale. Image courtesy of the author. **b** Picture taken in 2018 of the building of the Centre de Recherche Interdisciplinaire, in Paris, during the Nuit Blanche. It shows the projection on the original building of an image taken on a microscope of the self-organisation of actin filaments (red) and myosins (green) on a miniaturised building at the micron scale. Image courtesy of the author. **c** Picture taken in 2018 of the building of the Centre de Recherche Interdisciplinaire, in Paris, during the Nuit Blanche. It shows the projection on the original building of an image taken on a microscope of the self-organisation of actin filaments on a miniaturised building at the micron scale. Image courtesy of the author

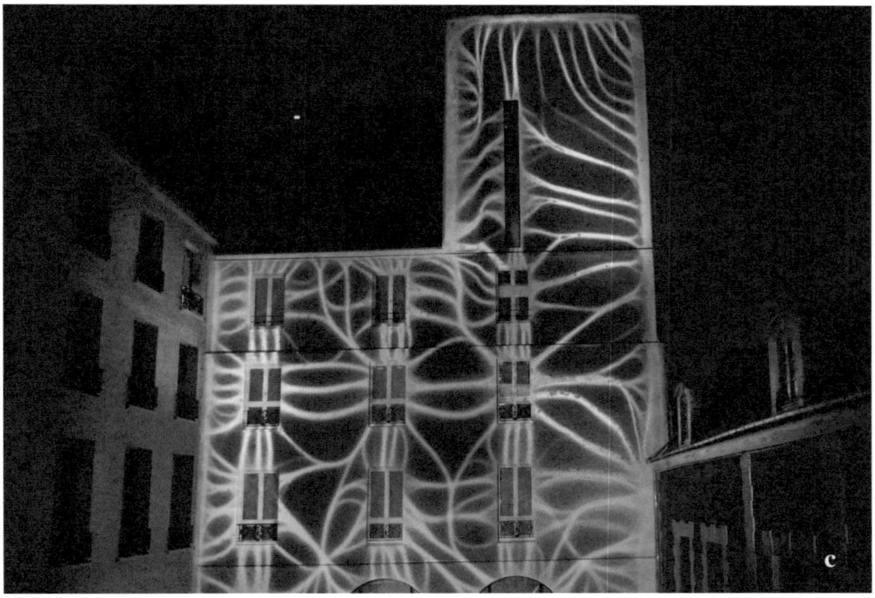

Fig. 4.20 (continued)

4.3.21 *Labor.* Paul Vanouse, 2019

By Paul Vanouse

What does exploitation smell like? *Labor* (Fig. 4.21) is an art installation that re-creates the scent of people exerting themselves under stressful conditions. There are, however, no people involved in making the smell, as it is created by bacteria propagating in the three bioreactors in the artwork. Each bioreactor incubates a unique species of human skin bacteria responsible for the primary scent of sweating bodies: *staphylococcus epidermidis, corynebacterium xerosis* and *propionibacterium avidum*. As these bacteria metabolise sugars and fats, they create the distinct smells associated with human exertion, stress and anxiety. Their scents combine in the central chamber in which a sweatshop icon, a wearer-less white T-shirt, is infused as the scents disseminate out. Scents intensify throughout the exhibition (Vanouse 2019).

The *Labor* installation is a post-anthropocentric scientific theatre in which audiences experience a microbial performance and simultaneously a notable absence. In the middle of the room, and at the absolute centre of the installation, connected to a radiating network of tubes, cables and bioreactors is an empty, white t-shirt hanging in a glass bell jar. While humans were at the centre of our world view, our ontology, our epistemology, in *Labor* the human has conspicuously vanished. Like Robert Boyle's bell jar, at the centre of an experimental proof of the existence of a vacuum and the icon of an emerging scientific episteme in the 17th Century (Shapin & Schaffer 2011), this bell jar rests at the centre of a new, tentative worldview, in which humans have vacated the centre (Vanouse 2020).

The smells of human body odours are generally abject, unwanted and impolite. Much of Western culture has been designed, constructed and marketed, to minimise such odours. Deodorants, toothpastes, mouthwash, breath mints, laundry detergents, soaps, room deodorizers, not to mention bathrooms and the modern toilet have been equated with both civility and civilization itself. Anne McClintock describes European soap fetishism, in particular, as "central to industrial modernity, inhabiting and mediating the uncertain threshold zones between domesticity and industry, metropolis and empire"

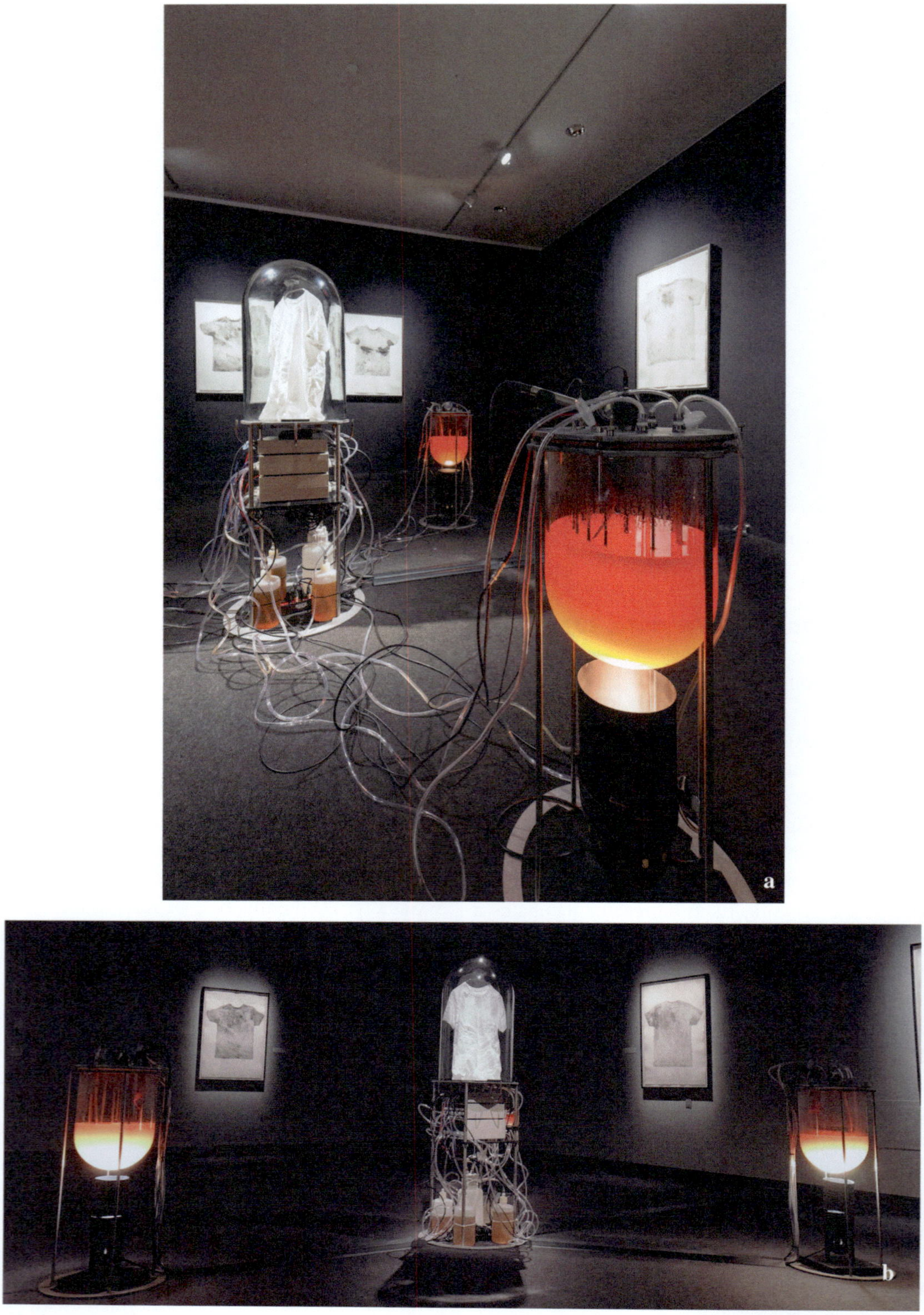

Fig. 4.21 Paul Vanouse. 2019. *Labor*. Multi-sensory installation. Burchfield Penny Art Gallery, Buffalo, NY. **a** Prix Ars Electronica Festival, OK Center, Linz, Austria. Photo: Otto Saxinger. **b** General view. Photo: Francesca Bond/Buffalo News. **c** General view. Photo: Tullis Johnson. **d** Detail. Photo: Tullis Johnson. **e** *Martin, Front*. Sweat-stain-transfer. Print. Photo: Tullis Johnson

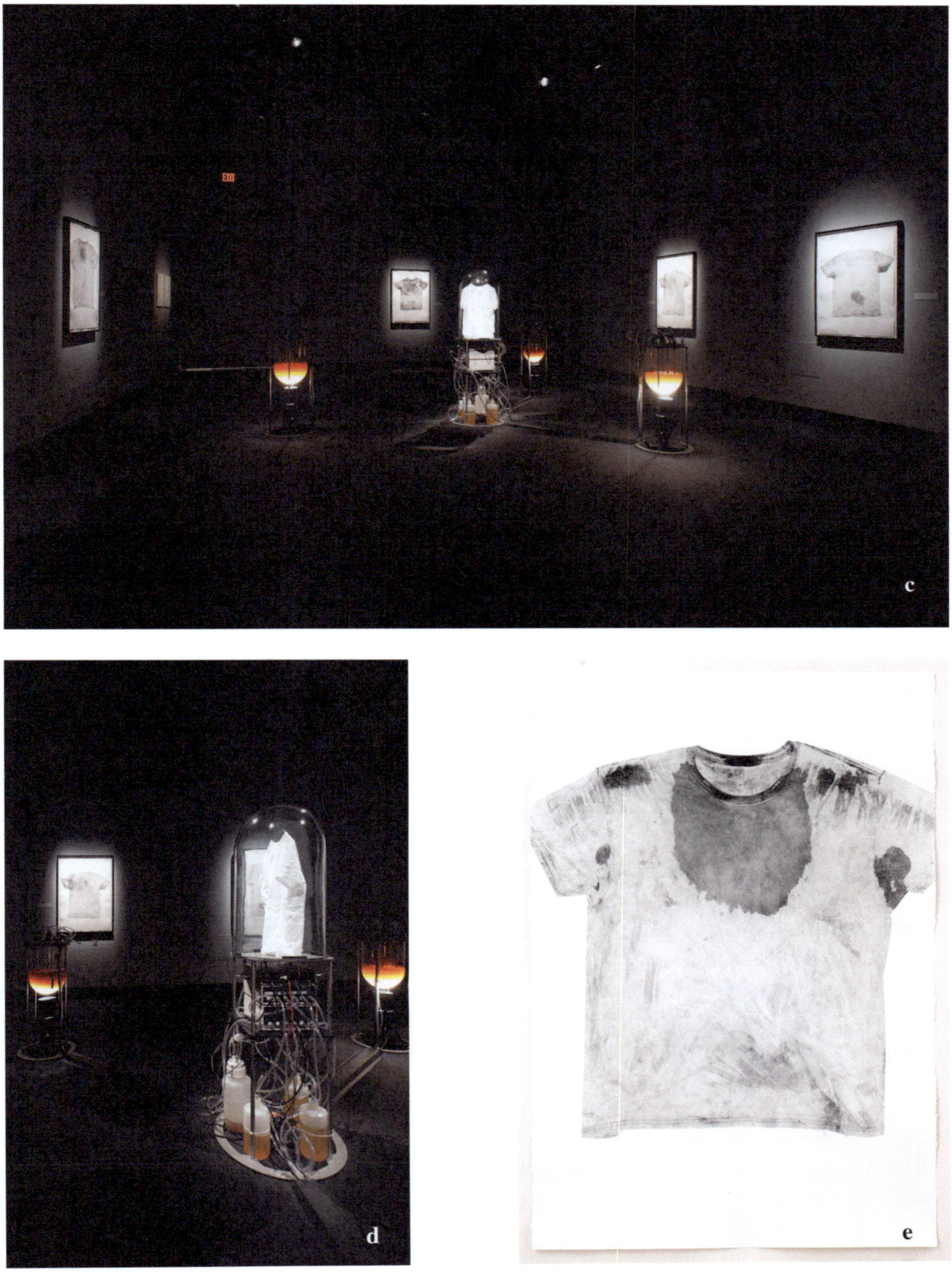

Fig. 4.21 (continued)

(McClintock 1995, p 210). Odour stories and myths of their effects, have well served racism, colonialism, imperialism and classism, reinforcing stereotypes, enacting boundaries and dehumanising perceived others. By artificially recreating human body odour, de-corporealized and in excess, and offering it for aesthetic contemplation I hope to befoul cultural prejudices through which both the faculty of smell as well as the quality of having a smell are denigrated. But more importantly, I wanted to create an apparatus for re-imagining relationships; a system that de-contextualizes the scent from its stigma.

I initiated this project at the Biofilia laboratory in Helsinki in 2014. There I isolated *staphylococcus epidermidis*, which produces a mild scent, and *propionibacteria*, which produces funky, acrid odours. I isolated them by covering my armpits in sterile pads in a day that included mild exercise and white-collar labour: writing proposals, replying to emails and scheduling meetings.

Cultivating skin bacteria in vitro is difficult as these strains are best suited to the complex ecosystem of human skin. At the Coalesce lab at the University of Buffalo, lab manager Solon Morse and I built custom incubation equipment to investigate at which metabolic stages, and under which conditions these organisms are most fragrant. I have not engineered genes into microbial production strains (as a synthetic biology approach would entail), but rather adjusted incubation conditions to fit the organisms and maximize their particular scents. The installation design incorporates stirrers, biosensors, heaters and electronic valves to automate these conditions. This was conceptually important, in a Foucauldian, biopolitical sense, as the control not just of labouring bodies, but fully integrated throughout the process of reproduction, growth and death.

The six sweat-stain-transfer prints capture visual evidence of exertion by employing a novel production method. Moist, freshly soiled shirts are dusted with charcoal and pressed between paper at high pressures, which infuses the paper surface with their image. Each T-shirt was worn by a university student paid at the prevailing wage.

I developed this method at Cultivamos Cultura in Portugal. After heavy labouring, I collected the sweaty T-shirts of the residents and I dusted them with charcoal, sandwiched them between paper and boards, and ran them over with an SUV. This produces embossed evidence of labour, portraits not of people nor of shirts but patterns of intimate excretion. Charles Sanders Pierce has defined an *indexical* component in a tri-part sign as a direct connection to the materiality of an object. In this sense, each image is an indexical imprint of an individual's sweating body (Pierce 1938).

The idea of what exactly makes up a person is again under the microscope, not just for scientists, but for culture at large. For centuries, we have debated who gets to be considered a person and when: it is a question that dominates political discourse of the last few centuries because of the connection to labour and liberation in the post-Renaissance world (Vanouse 2019).

Labour reflects upon our changing understanding of what we are. Microbes in and on the human body vastly outnumber human cells and they help regulate many of our bodily processes, from digestive and immune systems to emotional and physiological responses like sweating. Our microbiota is integral to who and what we are and complicates any simplistic sense of self. Likewise, the smell of the perspiring body is not just a human scent, unless we are willing to redefine what we mean by human. (Vanouse 2019).

The *Labor* project also reflects upon an industrial shift from human and machine labour to increasingly pervasive forms of microbial manufacturing. Today, microbes produce a wide range of products, including enzymes, foods, feedstocks, fuels and pharmaceuticals. They literally live to work. These new industrial activities point to a deepening of the exploitation of life and living processes: the design, engineering, management and commodification of life itself. In *Labor*, the microorganisms ironically produce the scent of sweat, not as a vulgar *bi-product* of production, like in factories of the nineteenth and twentieth centuries, but as a nostalgic *end-product* (Vanouse 2019).

As audiences contemplate clues within the installation, they will hopefully ponder perverse aspects of contemporary production like life inseparable from labour, product and produce conflated, biology a subfield of technology, and "man" omitted from manufacturing (Vanouse 2019).

References

McClintock A (1995) *Imperial Leather*, Routledge Press

Peirce C S (1938) Collected Papers of Charles Sanders Peirce, vols 1–6. Harvard University Press

Vanouse P (2019) *Labor*. Retrieved from http://www.paulvanouse.com/labor.html. Reviewed 4 February 2023

Vanouse P (2020) *Labor:* The Post-Anthropocentric Body 'at Work'. *Performance Research, 25*(3). Routledge Press

4.3.22 *Neuroportraits.* Nicholas Wade, 2011–2012

By Nicholas Wade

Neuroportraits (https://neuroportraits.eu) is a website linking art and the history of neuroscience (Fig. 4.22). Rather than presenting conventional portraits of significant figures from the past derived from engravings or paintings, they are combined graphically with a motif that captures the significance of their work. These 'perceptual portraits' are presented with the aim of combining art, history and science. Perceptual portraits generally consist of at least two elements—the portrait and some appropriate motif. The nature of the latter depends upon the endeavours for which the portrayed person was known. In some cases, the motif is drawn specifically to display a phenomenon associated with the individual, while in others it is derived from a figure or text in one of their books. The portraits and motifs have themselves been manipulated in a variety of ways, using graphic, photographic, and computer graphic procedures. Four illustrations here are derived from almost 200 perceptual portraits in https://neuroportraits.eu and they cover a period of more than five centuries of scientific advancements. The fifth illustration is a homage to Magritte's painting *The perfidy of images* (1929).

Figure 4.22a portrays Andreas Vesalius (1514–1564) and Thomas Willis (1621–1675). The renaissance of anatomy is considered to have been heralded by Vesalius in his *De Humani Corporis Fabrica.* The prints derived from Vesalius's dissections are taken to be a synthesis of science and art. The perceptual portrait of Vesalius is presented within the woodcut of the brain as seen from below (from *Fabrica*). Thomas Willis and his circle of anatomists at Oxford pursued the examination of the brain rather than relying solely on the texts written by others. His *Cerebri Anatome* placed dissections of the brain on firmer foundations: he numbered the cranial nerves and described the arterial circle at the base of the brain which still bears his name. In Willis' images the anatomy of brain circumvolutions and sulci are more clearly defined than in the representations by Vesalius. Willis is shown enclosed within the diagram of the brain (drawn by Christopher Wren) taken from his *Cerebri Anatome*, together with its title page.

Descriptions of the gross structure of the brain were based on dissection and direct observation. Detailed anatomy required instruments that would extend the range of human vision. Microscopes were introduced in the 17th century but they were subject to various optical aberrations. These were overcome in the eighteenth and nineteenth centuries when achromatic lenses were incorporated in microscopes. Difficulties still remained due to the seemingly transparent cellular structures. A huge advance followed from chemically staining the specimens so that nerve cells could be seen in more detail. The reazione nera (black reaction) devised by Camillo Golgi (1843–1926) in 1873 consisted of hardening the preparation in potassium bichromate and then impregnating it with silver nitrate. It exposed networks of nerves in

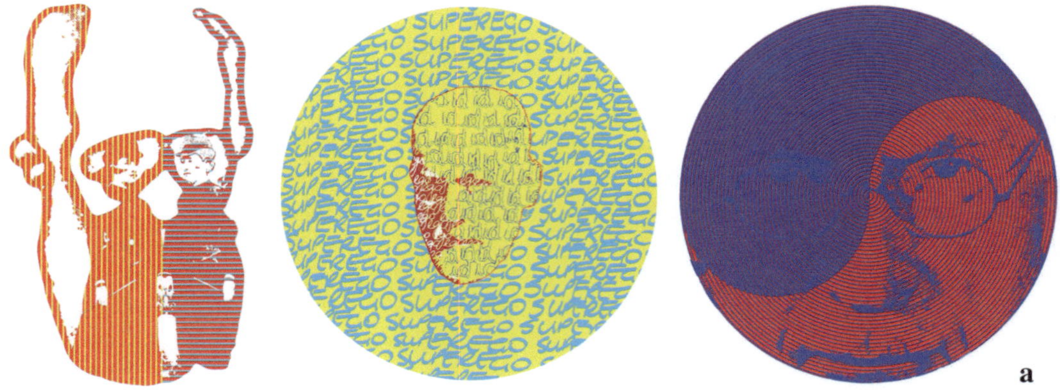

a

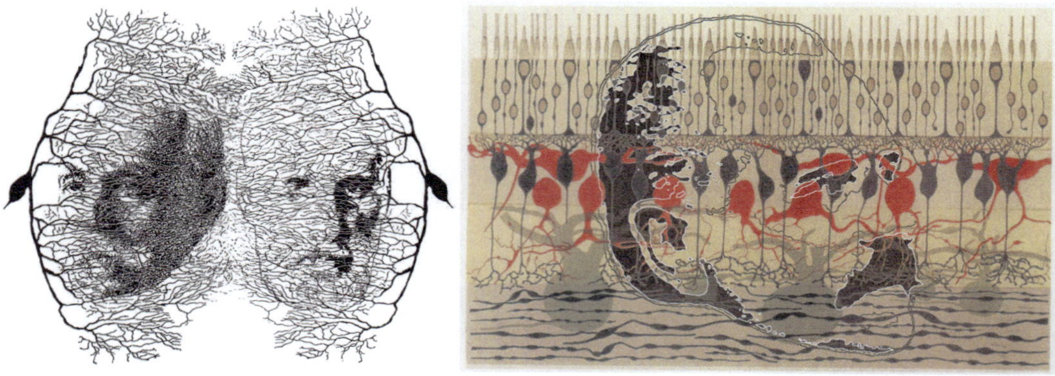

b

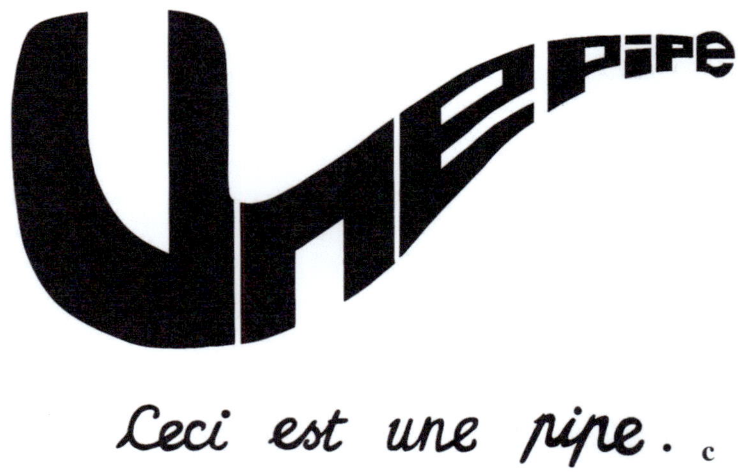

c

Fig. 4.22 **a** Nicholas Wade. 2018. *Explorers of the mind—Breuer, Freud and Jung*. Web page graphics. Courtesy of the author. **b** Nicholas Wade. 2018. *Golgi and Cajal in Purkinje cells and Tartuferi in his retina*. Web page graphics. Image courtesy of the author. **c** Nicholas Wade. 2018. The integrity of words. **d** Nicholas Wade. 2018. The brains of Vesalius and Willis. **e** Nicholas Wade. 2018. *Nativist (Descartes) and empiricist (Locke)*. Web page graphics. Courtesy of the author

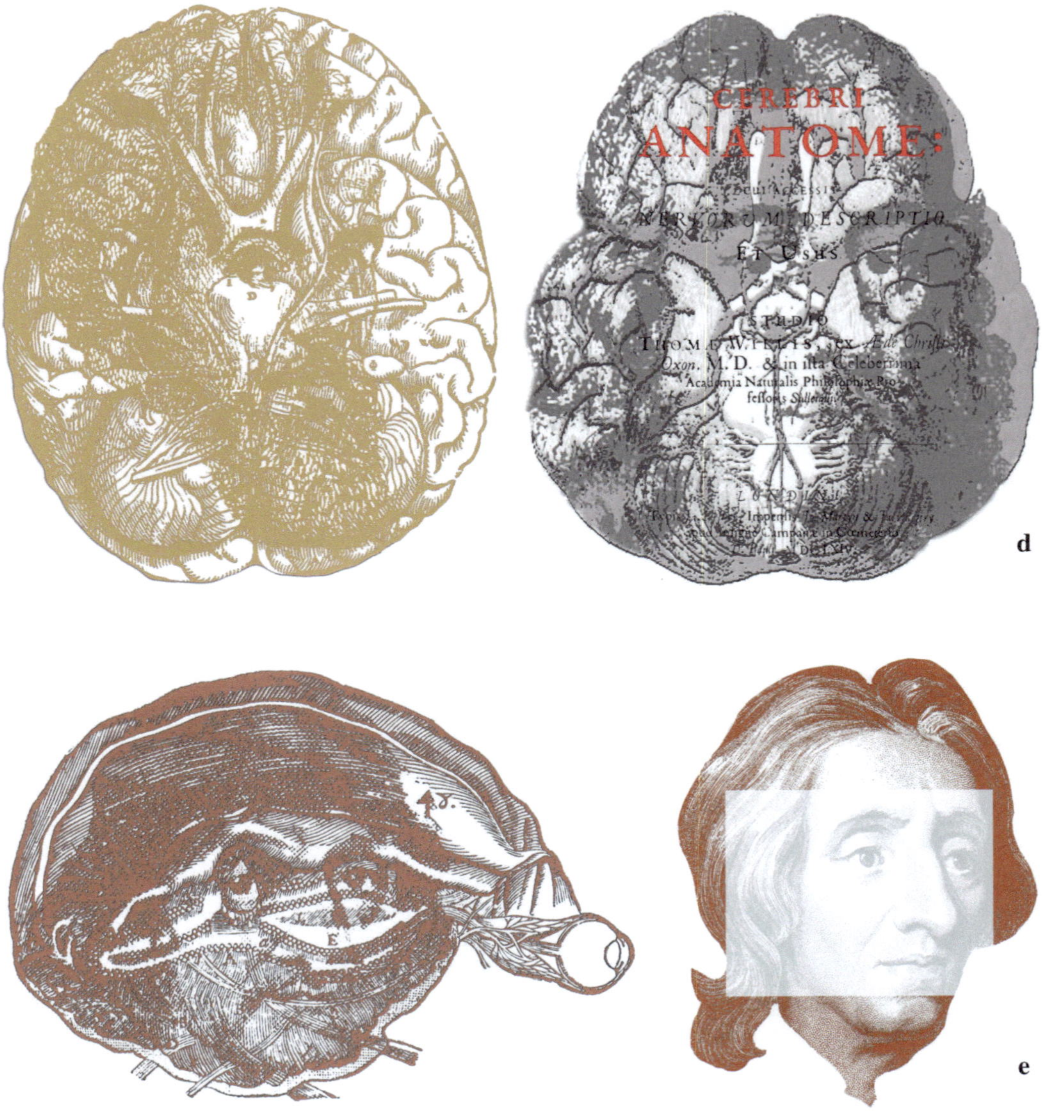

Fig. 4.22 (continued)

grey matter in a manner that had not been possible previously. Golgi is shown in his diagram of a Purkinje cell and its complex arborizations in Fig. 4.22b, as is Santiago Ramón y Cajal (1852–1934). Golgi's black reaction transformed histological studies of the nervous system. Cajal adopted and adapted it with great skill and his detailed drawings were examples of scientific art. It was one of Golgi's students, Ferrucio Tartuferi (1852–1925), who first applied the silver staining technique to the retina, and his portrait is contained in his beautiful illustration of the retina (Fig. 4.22b).

The early neuroscientists influenced the ideas of philosophers and would in turn interpret their results in terms of the then contemporary philosophies. Two prominent and opposing ideas were expressed by René Descartes (1596–1650) and John Locke (1632–1704) who are both portrayed perceptually in Fig. 4.22c. Descartes presented a nativist theory of mind whereas Locke considered that all knowledge was learned.

Locke rejected the Cartesian rationalism prevailing in continental Europe and established empiricism in Britain.

Some neurologists had a great influence on art when they ventured into the territory of the mind. This was evident from the work of Josef Breuer (1842–1925), Sigmund Freud (1856–1939) and Carl Gustav Jung (1875–1961) who are shown in Fig. 4.22d. Breuer interested Freud in hysteria by discussing one of his cases with him. Breuer is depicted in combination with two views of Anna O reflecting her dual personality. Freud set in train a major change in theories of neuroses. He established psychoanalysis as a treatment and developed a theory of mind that emphasised unconscious motivation. Three forces and their interactions fashioned behaviour: the *id* represents unconscious, biological desires which are partially repressed by the *superego*; a resolution is achieved by the preconscious *ego*, which also interacts via the senses with the real world. In contrast, Jung established an analytic psychology that emphasised the self. He was later struck by the symbols and myths used in all human cultures. In particular, the mandala was treated as a symbol of total unity. These included yin-yang representing the union of opposites, and Jung is shown within a variant of this symbol made up of concentric circles.

Magritte painted a picture representing a pipe beneath which were the painted words 'This is not a pipe'. The title of the painting is *The perfidy of images.* He is arguing (in paint) that neither the spatial representation of a specific pipe nor the symbolic word signifying the class of pipes are adequate representations of the object and that they are in conflict. My response to this is shown in Fig. 4.22e and entitled *The integrity of words.* In this case both the spatialized and stylized images are in harmony rather than conflict.

References (with Illustrations) for Further Reading

Wade N (1982) *The Art and Science of Visual Illusions*. London: Routledge & Kegan Paul.

Wade N (1987) "Allusory contours". In Petry S and Meyer G (eds.) *The Perception of Illusory Contours*. Springer, New York, 1987, pp. 274–294.

Wade N (1990) *Visual Allusions: Pictures of Perception*. London: Lawrence Erlbaum.

Wade N (1995) *Psychologists in Word and Image*. Cambridge, MA: MIT Press.

Wade N (1995) "Portraits of artists and scientists". In Gregory R L ,Harris J, Heard P and Rose D (eds.) *The Artful Eye*. Oxford University Press, Oxford, pp. 399–425.

Wade N (2007) *Circles: Science, Sense and Symbol*. Dundee: Dundee University Press.

Wade N J (2009) "Allusions to visual representation". In Skov M and Vartanian O (eds.) *Neuroaesthetics*. Amityville NY: Baywood Publishing. pp. 171–220.

Wade N (2016) *Art and Illusionists*. Heidelberg: Springer.

Wade N J (2017) "Hidden images". In Shapiro A and Todorovic D (eds.) *Oxford Compendium of Visual Illusions*. Oxford: Oxford University Press. pp. 778–784

Wade N J (2020) "Art and neurologists". In Colombo B (ed.) *Brain and Art: From Aesthetics to Therapeutics*. New York: Springer. pp. 53–66.

Wade N J (2021) On stereoscopic art. *i-Perception*, 12(3), 1–17. https://doi.org/10.1177/20416695211007146.

Wade N J (2021) On the art of binocular rivalry. *i-Perception*, 12(6), 1–23. https://doi.org/10.1177/20416695211053877

Wade N (2023). *Vision and Art with Two Eyes*. Heidelberg: Springer.

4.4 Biographies

Pierre Berger. Mathematician researcher at the Centre National de la Recherche Scientifique (CNRS) and the University of Paris 13, and multimedia artist. He has created and coordinated numerous exhibitions in Paris, Rio de Janeiro, Mexico and Berlin in which he confronts spaces of the mathematical imagination to capture the feelings they produce. His artistic work is nourished by new advances in basic science and is expressed through multimedia format. His training encompasses science and art: he was a student at the prestigious École Normale

Supérieure and was trained at the French École Nationale Supérieure des Arts Décoratifs.

Vincent Martial. Artist, musician, composer, Vincent Martial focuses his work on sound, image and movement, as well as experimentation to create transmissive forms. He is interested in live sound and movement created within stage performances, as well as the sound environment, the attention we pay to it and the use of it within the artistic domain. Much of his work involves creating devices that serve as frames for performances, installations, shows or exhibitions. His works have been presented in Europe, Asia, North and Central America, in places such as Tate Modern London, Guggenheim New York, Bethlehem Cultural Center (Lisbon), Roman Theater (Lisbon Archaeological Museum), Césaré CNCM, le Cube, La Comédie de Reims, TJP Strasbourg, TNG Lyon, Lieu Multiple, l'Aeronef, Festival Big Bang and Jazzmandu, among others. His work *Pipe-Koto* #2 belongs to the Tate Modern collection. He is also a composer for theater, puppet and dance companies such as L'Ateuchus Succursale101, Naforo-ba, Théâtre de l'estrade.

Jimena Royo-Letelier. Chilean artist and researcher who arrived in France in 2009 to study at the École Polytechnique, obtaining a doctorate degree in applied mathematics at Université de Versailles Saint-Quentin-en-Yvelines. Graduated from IRCAM and Université Pierre-et-Marie-Curie Paris-VI, Master from the École Normale Supérieure de Cachan, she divides her time between teaching at the University of Santiago Chile, researching in computational music and in artistic projects that bring together sound, mathematics and socio-political themes. In 2016 she created the Lakeri collective with the composer Alice Guerlot-Kourouklis, producing the sound and visual installation *Murs Invisibles*, a work that has been presented in France, Canada, Colombia and Chile. In 2016, together with Pierre Berger and Sergio Krakowski, she created the interactive sound piece Conversations, presented at the 7th Berlin Mathematics Congress and is now part of the permanent collection of Universum Museum of Science and Technology, Mexico City. Since 2017, together with Berger and Vincent Martial, she has created several sound and plastic sculptures for the art and science research group "Esthétopies", which works on the sensitive exploration of mathematical spaces.

Jean-Baptiste Boulé. Ph.D in Biochemistry and Molecular Biology. Principal investigator at the Centre National de la Recherche Scientifique (CNRS), he was Director of the Genome Structure and Instability Laboratory. He is the head of the Genomics and Physiology of Adaptation team at the National Museum of Natural History and participated in the Research Division of the Institute Curie Unit Dynamics of Genetic Information: Fundamental Bases and Cancer.

Nicolas Desprat. With a degree in fundamental physics from the Pierre et Marie Curie University, he received his doctorate in condensed matter and is presently serving as a professor at the University of Paris. His research focuses on understanding how physical limitations shape microbial dynamics in Bacterial communities, Host-pathogen interactions, Collective behaviours in *Chlamydomonas reinhardtii* and *Saccharomyces cerevisiae* in fluctuating environments, and Evolution in a spatially structured environment.

Philippe Parreno. Born in Algeria, he is a French contemporary artist who lives and works in Paris. His works include films, installations, performances, drawings and texts. Parreno's work focuses both on the expansion of ideas of time and duration and on the distinctive conception of exhibitions as a medium. He began examining unique approaches to both storytelling and representation in the 1990s, and has since exhibited internationally. His works are part of the collections of many institutions such as the Centre Georges Pompidou, France; 21st Century Museum of Contemporary Art,

Kanazawa Japan and the Museum of Modern Art New York.

Dmitry Bulatov. Artist, art theorist and curator at the State Center of Contemporary Art within the Baltic branch of the Pushkin State Museum of Fine Arts. He has spearheaded numerous exhibition and publishing ventures spanning the realm of new media art and scientific inquiry. His artistic contributions have graced prestigious platforms such as the 49th and 50th Venice Biennale (2001, 2003) and Ars Electronica (Austria) in 2002. Dmitry is an accomplished author, penning books and anthologies on contemporary art, and holds editorial positions at esteemed magazines including DOC(K)S (France) and NOEMA (Italy). He has won the State Prize "Innovation" twice in the field of Contemporary Art (Russia 2008, 2013). In 2014 he was nominated for the Golden Nica at the Prix Ars Electronica festival in the category of Visionary Pioneers of Media Art.

Alexey Chebykin. Independent conceptual artist, architect, and multidisciplinary multimedia artist established in Kaliningrad, works at the intersection of contemporary art and new technologies. His interest is focused on Digital Art and Generative Art. He has participated in many exhibitions including The Heart of Venice (Italy, 2007), PRAGUEBIENNALE 3 (Prague, 2007), 52th Venetian biennial (GINNUNGAGAP/ Pavilion of Belief, 2007), Artissima (Turin, 2007), Schengen (Berlin, 2008), Russian Utopias (Moscow, 2010), Life.Scientific Version (Moscow, 2011) and the Venice and Prague Art Biennials.

Juan M. Castro. Artist and researcher in soft matter and systems chemistry based in Japan. He works mainly in hybrid installations on themes centred on the origin or origins of life, material agency, deep space and extraterrestrial life. His work employs a variety of media, such as extraterrestrial organic matter, lipids, amino acids, genetic material, protocrystalline genetic material, protocell models, microbes, high-pressure and high-temperature reactors and computer graphics. He has presented his projects internationally in museums, galleries and multimedia art festivals, such us the Science Gallery Ireland; the NTT InterCommunication Center [ICC] Japan; the Central House of Artists Russia; the Museum of Applied Arts and Design Estonia; the Museum of Applied Museum of Applied Arts and Design Estonia; the Esther Klein Gallery-Science Center USA, among others.

Akihiro Kubota. Member of the PЯOTO-ALIEИ PЯOTOJECT (PЯOTO-A), Professor of Art and Media at the Department of Information Design, Tama Art University, Tokyo. He received his doctoral degree from the School of Engineering, University of Tokyo. He has pioneered the development and synthesis of a wide spectrum of hybrid creative initiatives, such as satellite art, bio-art, digital fabrication and sound performance with instruments invented by himself. His team's ARTSAT1:INVADER won the Prix Ars Electronica 2015 Hybrid Art Award of Distinction. He received the 66th Ministry of Education Award for Fine Arts (Audiovisual Arts Division) for the achievements of the ARTSAT project.

Taro Toyota. He is part of the PЯOTO-ALIEИ PЯOJECT (PЯOTO-A). Chemist by training, he works at the Department of Basic Science, Graduate School of Arts and Sciences, University of Tokyo. He has received several grants for his research including the Ministry of Education, Culture, Sports, Science and Technology; Japan Society for the Promotion of Science and the Japan Science and Technology Agency.

Shanthi Chandrasekar. Multidisciplinary and multimedia artist with academic background in Physics and Psychology. Her interest in understanding different visual media has led her to experiment with sculpture, photography, drawing, painting, engraving and paper. She has also been trained in the traditional art forms of Kolam and Tanjore. Deeply influenced by her Indian heritage, she finds inspiration from the enigmatic magnificence of the world around her where the scientific overlaps with the spiritual. She has won Individual artist awards and fellowships from the Maryland State Arts Council and the Arts and Humanities Council of Montgomery County, as well as the Maryland Traditions Master Apprentice Award for teaching kolam. Shanthi has exhibited and presented her work at various galleries and institutions, including the Fermi National Accelerator Laboratory, the Gandhi Memorial Center, and the Smithsonian.

Matthias Oostrik and Suzanne Dikker. Their collaborative work explores one of the most fundamental questions of human nature, which lies at the very core of artistic, scientific and technological research: what does it mean to be "connected" to another person? DIKKER + OOSTRIK are widely recognized as pioneers in both the fields of biofeedback art and real-world crowdsourcing neuroscience. Their work (including *Measuring the Magic of Mutual Gaze* with performance artist Marina Abramovic, *Ondas* with musical performers Residente and Bad Bunny; and *Mutual Wave Machine* with video artist and Guggenheim Fellow Peter Burr) is supported by the Dutch Fund for Creative Industries, Niemeijer Fonds, The Max Planck Society, Hersenstichting and the Netherlands Organization for Scientific Research, among others. It has been exhibited in several museums and festivals, including the Garage Museum for Contemporary Culture (Moscow), the Benaki Museum (Athens), the American Museum of Natural History (New York), the EYE Amsterdam (Amsterdam), the National Ballet of Marseille (Marseille), the Deutsches Filmmuseum (Frankfurt), Espacio Telefónica (Madrid), Lowlands (Netherlands), the National Museum of Modern Art (MNAC, Lisbon) and the De Lakenhal Museum (Leiden, Netherlands). Suzanne is an associate research professor at New York University, and Matthias Oostrik is an independent artist.

Suzan Tunca. Head of the ICK Academy, artistic research coach at Codarts University of the Arts, and Ph.D candidate at ACPA Leiden University. As a dancer, dance researcher, teacher and choreographer she supports the regeneration and advancement of the art of dance and its capacity to embody invaluable resources of knowledge and understanding. ICK (International Choreographic Arts Centre), directed by Emio Greco en Pieter C. Scholten, is an acclaimed dance platform that produces international dance performances, guides guest artists and initiates various education and research projects. The company received various prestigious prizes and toured all over the world during the past twenty years.

Than van Nispen. Composer van Nispen is a lecturer and researcher at HKU School of Music and Technology and the Expertise Centre for Creative Technologies. Than is also a music composer for games and interaction at the GreenCouch composition and sound design collective.

Diederik Schoorl. Creative Technician Diederik Schoorl Spectro Productions has a long track record of creating technology for art and artists, such as Joris Strijbos, Nicky Assmann and Frank Plant. For DIKKER + OOSTRIK he has designed and created the technology behind the Mutual Wave Machine and Harmonic Dissonance.

Gilberto Esparza. He investigates technology as a means to raise questions and solutions regarding the impacts of the human footprint on life on Earth. His work is based on a vindication of the intelligence inherent in life and a rethinking of the relationship between human societies and the natural environment. His practice employs recycling consumer technology and experiments with biotechnology, collaborating with research centres such as the Chemical and Process Engineering Group of the University of Cartagena, Spain; Mechatronics Area of the CINVESTAV of the National Polytechnic Institute, Institute of Engineering, Juriquilla, UNAM; or the Digital Arts, University of Guanajuato, Mexico. He has held individual and group exhibitions in Mexico, the United States,

Canada, Brazil, Colombia, Peru, Ecuador, Argentina, Spain, Holland, Belgium, Slovenia and Catar. He received the Golden Nika, Hybrid Art at the Prix Ars Electronica 2015; second prize Vida 13 from Fundación Telefónica, Spain; honourable mention in Ars Electrónica, Linz Austria; and the Ibero-American Production award VIDA 09. He is currently a member of the National System of Creators of Art in Mexico.

Kathy High. She is a professor of Video and New Media in the Arts department at Rensselaer Polytechnic Institute and coordinator of the NATURE Lab at the community organisation Sanctuary for Independent, Troy Media (New York). Artist and educator, she collaborates with scientists and other professionals working on thematic areas as Bio-Art, Animal/Interspecies, Science Fiction, Queer/Body, Activism/Antiwar, Animal sentience and Ethical dilemmas of biotechnology and the medical industry. Co-author, with Sherry Miller Hocking and Mona Jimenez of the book "The Emergence of Video Processing Tools" and co-editor, with Ralph Hocking and Sherry Miller Hocking of "Experimental Television Center 1969-200"9. She also edited *FELIX*: *The Journal of Media Arts & Communication during* 1991-2003.

David Goodsell. He earned a bachelor's degree in biology and chemistry at the University of California, Irvine, and completed a Ph.D in DNA X-ray crystallography at the University of California, Los Angeles, in 1987. He is currently an illustrator and Professor of Computational Biology at the Scripps Research Institute and a Research Professor at Rutgers University, where he directs scientific dissemination of the RCSB Protein Data Bank. In 2022, he was honoured with the Carl Brändén Award from The Protein Society, and the I. Fankuchen Award from the American Crystallographic Association. In 2009 he published the book "The Machinery of Life", which became a best-selling book. During the years, he has written more that two hundred papers, which can be accessed via the National Library of Biology for the United State, and most of his drawing can be downloaded for free, under Licence Creative Commons.

Yunchul Kim. Dynamically integrating science, technology, music, philosophy and poetry into his practice, Kim unravels a universe of material entanglements. His trans-material work has been recognized with the Collide International Award 2016 (CERN), and awards at VIDA 15.0, Ars Electronica and Transmediale, among others. Actively engaged in cutting-edge collaborative research ventures like *Mattereality and Liquid* Things at the University of Applied Arts in Vienna, and the art and science project group Fluid Skies. In 2014, Kim established Studio Locus Solus in Seoul. His artistic endeavours have been showcased globally, including exhibitions at the Yokohama Triennial in Japan; KUMU in Tallinn; CCCB in Barcelona; FACT in Liverpool; ZKM in Germany; Ars Electronica in Austria; the International Triennial of New Media Art in China; VIDA 15.0 in Spain; Transmediale in Germany; ISEA in Germany; and the New York Digital Salon, among others. His solo exhibition *Gyre* was on display at the Korean Pavilion during the 2022 Venice Biennale.

François-Joseph Lapointe. Artist and scientist at the University of Montreal. He has a doctorate in evolutionary biology (1992) and a second doctorate in dance and performance (2012). As part of his scientific research, he has published more than 120 articles in the fields of molecular systematics, population genetics and metagenomics. As part of his artistic practice, he applies biotechnology for creative purposes and has created *choreogenetics*: the generation of choreographic variants through genetic mutations and selection. His research has been funded by the National Science Foundation (USA), Genome Canada, the Natural Sciences and Engineering Research Council (NSERC), the Social Sciences and Humanities Research Council (SSHRC), the Fonds de recherche du Québec - Nature et Technologies (FRQNT) and the Fonds de recherche du Québec - Société et Culture (FRQSC). As a bio-performance artist, he has been exhibited at several institutions including the Medical Museion (Copenhagen), Science Gallery (London), the Musée de la civilisation (Quebec), Transmediale (Berlin), SciArt Center (New York), Ars Electronica (Linz), and the Centre Pompidou (Paris).

Marianne Cloutier. She works with living materials and biotechnologies, speculative biology and representations of the body and motherhood, with special interest in epistemology and research-creation methodologies. She earned a doctorate in Art History, specialising in the question of identity in bioart. She was co-investigator of the project "La conduite responsable en recherche-création: outiller de façon créative pour répondre aux enjeux d'une pratique en effervescent" (Responsible management in research-creation: creative tools to meet the challenges of an effervescent practice), a concerted action of the Fonds de recherche du Québec. She was the co-author of a guide entitled "Trousse d'accompagnement en CRRC : synthèse des enjeux de conduite responsable en recherche-création et proposition d'outils réflexifs" (Toolbox in RCRC: synthesis of responsibly conducted research-creation issues and proposed tools for reflection). In addition she was co-curator of the exhibition Art+ Bioethics (2016), and worked since 2019 as coordinator and curator of the McConnell-University of Montreal Chair on the Reappropriation of Motherhood. She is professor in the Department of Art History at UQAM and at the D.E.S.S. in Arts, Creation and Technologies from the University of Montreal, and external consultant to the CSSDM.

Max Liboiron. Holds a Ph.D in Media, Culture and Communication from New York University, with a dissertation titled "Plastics in the Wild", and a M.F.A. in Studio Art from the State University of New York at Stony Brook, complemented by an Advanced Graduate Certificate in Cultural Studies. Max's academic journey began with a B.F.A. with Distinction from Mount Allison University. Currently serving as an Associate Professor in the Department of Geography at Memorial University, Liboiron holds cross-appointments in Sociology and the Marine Institute Fisheries and Oceans program, with affiliations to the Program in Environmental Sciences. Noteworthy achievements include the project "Wild food movements and contaminants of concern in Imappivut" funded by POLAR Canada, and the creation of CLEAR, an interdisciplinary laboratory dedicated to tackling plastic pollution. Liboiron's recent initiatives include co-directing the IndigeLab Network, where Indigenous scholars collaborate on innovative methodologies to cultivate and support research communities. Alongside these academic pursuits, Max authored "Pollution is Colonialism" (Duke University Press 2021), and co-authored with Josh Lepawsky "Discard Studies: Wasting, Systems, and Power" (Cambridge: MIT Press 2022).

Rachel Mayeri. Los Angeles-based artist who works at the intersection of science and art. Her videos, installations and written projects explore themes from the history of special effects to the human animal. In 2011, she received a major arts grant from the UK-based Wellcome Trust to make original videos to entertain captive chimpanzees. The resulting project, commissioned by Arts Catalyst, has been exhibited at major film and arts festivals such as Sundance Berlinale, True/False Film Festival Missouri, Transitio México Electronic Art Festival, Abandon Normal Devices Manchester and Edinburgh Festival of Art, including Honorary Mention at the Hybrid Art Award at Ars Electronica. Her projects include *Primate Cinema: Baboons as Friends* (2007), *Stories from the Genome* (2003), and the compilation Soft Science (2004). Her most recent work is the multiscreen video installation *Life cycle of Toxoplasma Gondii* (2015), about the cycle of this parasite in rats, cats and humans. Her book chapter "Soft Science: Artists' Experiments in Documentary Storytelling" was published in "Tactical Biopolitics: Art, Activism, and Technoscience", edited by da Costa B. and Philip K. (MIT Press 2008). She is a visiting curator at the Museum of Jurassic Technology and a professor of Media Studies at Harvey Mudd College, Claremont.

Andro Montoya Riveros. He Ph.D in cellular biology and neurosciences, Mural and graffiti artist and Art and Science practitioner for exploration and learning. He is interested in developing and putting into practice teaching and oral communication skills in subjects related to biology, arts and nature. In relation to his academic training, the relation of neuroscience, immunology and wellbeing is of particular interest to him. Using his experience in Neuro-Immunology and Muralism, he contributes to the comprehensive understanding of living beings. Andro is an active participant of different graffiti communities in Chile, his work can be found in the open air throughout Chile and outside artistic institutions.

Špela Petrič. Slovenian artist with background in natural sciences and expertise in new media, she holds a Ph.D in Biology and in 2013 pursued a master's degree in Advanced Studies in Transmedia in 2013 with a thesis titled "Humalga: Towards the Human Spore". As a postdoctoral researcher at Smart Hybrid Forms, Vrije Universiteit Amsterdam, she blends biomedia and performativity in her artistic pursuits to explore unconventional connections between bodies and the foundations of our (bio)technological societies suggesting alternative narratives. Her exceptional artistic achievements include the Bioart and Design Award in the Netherlands, the 2022-23 Ars Electronica + The Art Collection Deutsche Telekom (Linz), and a Prix Ars Electronica Honorary Mention for her project PL'AI in 2021.

Perdita Phillips. Since 1992, her commitment to ecosystem thinking has led her to work with material and conceptual networks as diverse as drains, minerals, termites and birds, perched at the intersection of the human and non-human worlds. After studying Environmental Sciences, he was awarded with the British Commonwealth Scholarship to pursue a Master's degree at Goldsmiths College between 1997 and 1999. Phillips' field/hiking work (2003-2006) was recognized as one of the top three annual abstracts in the Leonardo Abstracts Service database. She has received two Australia Council Inter Arts scholarships and has participated in numerous exhibitions including: *Make Known: The Exquisite Order of Infinite Variation* (UNSW Galleries, Sydney 2018); *Frankenstein GRID* (Stanford University outdoor screening festival 2018); the Incinerator Art Award (Incinerator Gallery, Moonee Ponds 2017); Another Green World (Western Plains Cultural Centre, Dubbo 2017), *Objectillogica-a modern wunderkammer* (The Holmes à Court Gallery at Vasse Felix, 2017); and has participated in numerous interdisciplinary forums. Recent publications include *Fossil III* for the Lost Rocks project (2019).

Mikhail Rybakov. He is a multimedia artist and researcher who studied Media Art and Product Design at the Karlsruhe State University of Art and Design, Germany. From 2018 to 2020 he conducted postgraduate research on *"Arrival Strategies": designing spaces and interaction that make you feel present and embodied,* with professor Bruno Latour as advisor. He taught the product design course for first-year students at the Staatliche Hochschule für Gestaltung Karlsruhe, and stands out as a researcher and member of the "Critical Zones Study Group" of the exhibition at the ZKM of Critical Zones.

Susana Soares. Her work explores the design implications of the current technological redesign of nature and the future technological implications on public awareness and engagement in science. Her projects include developing collaborative frameworks between design and emerging scientific research. She is a senior lecturer at London South Bank University, researcher in the "IMPACT!" project from the Royal College of Art, and in "MATERIAL BELIEFS" from Goldsmiths University of London. Her current research focuses on how understanding technologically redesigned living systems can generate new frameworks for design practice. She has exhibited at the MoMA New York; MOMAK Kyoto; Science Gallery Dublin; the Southbank Centre and The Royal Institution, London.

Daina Tasmina. She is a Latvian mathematician currently Associate Professor at Cornell University. Known for creating crocheted objects to illustrate hyperbolic space, her book "Crocheting Adventures with Hyperbolic Planes" won the Bookseller/Diagram Award for Weirdest Title of the Year in 2009. In 2012 she was awarded with the Euler Prize from the American Mathematical Society. She has participated in exhibitions both in Europe and the United States. Her works of art are in the collections of several private collectors and universities, and have been included in the Smithsonian Museum's Collection of American Mathematical Models, the Cooper-Hewitt American Mathematical Models Collection of the Smithsonian Museum, the Cooper-Hewitt National Design Museum and the Henri Poincaré Institute in Paris.

Manuel Thery. His pioneering work supported the fact that cells respond not only to chemical signals, but also to the geometry of the environment. Théry's equations on the orientation axes of cell division represented an important advance. After joining the French Alternative Energies and Atomic Energy Commission (CEA) in 2006, he participated in the successful start-up Cytoo, dedicated to exploring the impact of microstructures on genome transcription and cell division. By establishing a laboratory branch at the Saint Louis Hospital in Paris, the Cytomorpholab aims to improve the control of blood cell cultures, with possible implications for the treatment of blood diseases. Théry's diverse contributions highlight his interdisciplinary approach and his commitment to advancing scientific knowledge in cell biology, deeply sensitised by art.

Paul Vanouse. He has an MFA from Carnegie Mellon University and BFA from University at Buffalo. Since the early 1990s, his works have addressed complex questions raised by various new technosciences. His works include data collection devices that examine the ramifications of surveys and categorization, genetic experiments that undermine scientific constructions of race and identity, and temporary organisations that playfully critique institutionalisation and corporatization. Paul Vanouse has presented his work in more than 100 spaces including *Topologies of the Real*: *Techne Shenzhen* 2023, Shenzhen Museum of Contemporary Art, China; *Gaia: Gene, algorithm, intelligent design, automata,* Museum of Contemporary Art, Taipei; and *The Measure of Chaos,* A. S. Popov Central Museum of Communications, St. Petersburg, Russia.

Nicholas Wade. British psychologist and academic, emeritus professor in the psychology department of the University of Dundee in Scotland. His research is focused on visual perception: the representation of space and movement in human vision; the history of research in visual sciences, in particular those related to stereoscopic vision; the relationship between visual science and visual art, illusions and the psychology of art and aesthetics. His objective is to foster a closer association between the graphic language of art, particularly geometric abstraction, and the interpretive language of science. He maintains an active international collaboration with Prof. S. Finger, from Washington University St. Louis and Dr. B. Lingelbach, Aalen University, among others. He has written and edited more than 15 books, including "Visual Allusions: Pictures of Perception" (London: Routledge 1990) and "Art and Illusionists"' (Heidelberg: Springer. 2016).

5 Capturing the Insights Behind the Selected Art-Science Works

Abstract

In this chapter, interviews conducted with most of the authors of the selected artist-science artworks were analysed and organised according to five categories. These categories emerged from their reflections on the premises, potential and limits of art-science integration. Their accounts point in the direction of a wide range of topics that are relevant to how humans think about themselves, its relation with other beings and entities, and complex global issues, amongst others. Their reflections also expose their view on the role of the arts and the sciences in society today.

Keywords

Art-scientists · interdisciplinarity · Artistic research · Science communication

In addition to presenting and describing in written format one of their artistic pieces (Chap. 4), interviews were carried out with at least one of the authors of each of the works based on a questionnaire, detailed in.[1] All of these occurred during 2021, except the interviews with Dmitry Bulatov, Gilberto Esparza, Rachel Mayeri, Perdita Phillips and Susana Soares which took place in 2016 using the same questionnaire.

To analyse the information from these interviews, the material was organised according to five categories: (1) motivation to engage in art-science; (2) scientific topics and lines of research; (3) materiality and artistic methodology; (4) differences between artistic and scientific approaches; and (5) art-science synthesis: Its meaning and implications. The participants provided clues about the differences between their artistic practice and scientific work, sometimes suggesting the limits and possibilities of each. Ideas alluding to the integration of art and science were also identified. In this way this research is both integrative and propositional. As readers will notice in what follows, the art-scientists reaches into diverse topics including finance, power, technology and interdisciplinarity, the relationship between artists and scientists and between them and the public. A picture of the roles of art and science in society today emerges.

---

[1] Questionnaire used in the interviews:

I. Regarding your artistic production, could you describe which subjects you have been working on over the last 2 years?
II. What makes your work "art"?
III. Which are the questions that have motivated your artistic work?

© The Author(s), under exclusive license to Springer Nature Switzerland AG 2024
M. Velasco and I. Nieto, *The Art-Science Symbiosis*, https://doi.org/10.1007/978-3-031-47404-0_5

5.1 Motivation to Engage in Art-Science

The art-scientist's motivations to undertake their works are often important to understand the artwork itself. Whilst each creator has his or her own personal motivations, some reasons were common to many of the artists-scientists.

(a) **For personal development**

The interviewees express different motivations for engaging in art-science according to their professional identity. On the one hand, the scientists wish to go beyond doing "just" science through experimentation with different types of materials, objects, and processes, and are motivated to communicate ideas that often have an ethical, social or political component (regardless of whether the final product is labelled science or art). Although the initial objective is producing the art-science piece, scientific publications are sometimes its offspring, which too is a kind of motivation. On the other hand, the motivation of artists is to work with experts from areas of knowledge that are new to them and become immersed in the world of science.

Both the artists and the scientists reveal their desire to walk the line between the limits of art and science. This drives and inspires them to find new methodologies and processes of creation. Some mention it is important to them that these be kinder to the earth than those of the past. Others point to scientific experiments that engage with new codes and visual forms of art.

(b) **Involvement in art or science**

Beyond personal development, the interviewees seek involvement in science or art as an exercise of collaboration and interdisciplinary work. Many remark that during the process of art-science integration, they remembered having had a prior interest or contact with art or science. The scientists express interest in ways that artistic experimentation might be usefully translated or transferred into their laboratories.

Scientists enjoy discovering the world of art as a healthy space that fosters the expression of feelings and emotions. Others find in it a meaningful space that indirectly challenges them to exploit the moments of secret poetry hidden in science. Like the postmodern artists of the 1980s who use the instruments of advertising to criticise capitalism and the mass market, art-scientists employ the means of science to question and criticise science itself.

Some of the art-scientists value this integration as bringing together people from different disciplines and a diversity of subjects. This openness promotes unprejudiced conversations and the ability to discuss and create without the pressure of having to take a stance, as usually occur when working with peers in specific disciplinary issues. They believe that art-science experiments are just as likely as science to impact the world and have the advantage of giving rise to novel ways of doing things, in a context where some aspects are familiar as some are not for both the artist and the scientist.

(c) **Engagement with new audiences**

Interviewees seek out new and different audiences they can connect to and inspire through their work. Some of the scientists

IV. From which disciplines (philosophical, scientific, artistic) do you start to develop your work?
V. How did you get to art? Do you have friends or relatives in art? Which references or sources do you explore?
VI. What is the difference between your scientific practice and your artistic work?
VII. In terms of methods, do you use a specific artistic influence or technique at some stage of your artistic work? If you do, what is its role?
VIII. Are you doing science when you are working on your artistic piece?
IX. Do you know any scientists who have used your results or methods to further their own research (beyond the conventional scientific framework)? Can you mention any examples?
X. Which audience is more comfortable to present your work? An artistic one? ¿A scientific one? Why?

report engaging with art to get closer to audiences by breaking out from under the requirement of "objective" science to remain detached from the object of study.

Becoming aware of the performative dimension of science is of value because it enables scientists to reach a wider audience than usual and to share with the general public the sheer *blast of pleasure* associated with scientific discovery. The aesthetics of the art-science performance can inspire people not related to the sciences, or to the specific subject considered, to learn on the subject or share their distanced, hence valuable, points of view.

Art-scientists express excitement that their work may be accepted by scientists and enjoyed by artists. Science can be isolating and repetitive which is what motivates some of the interviewees to work with people in contexts where engagement and connection are appreciated. Thus it is possible to be valued not only through the institutions of art or science, but through the mass media and in turn receiving recognition as a social actor.

(d) **As a way to raise and address complex questions**

Art-science integration opens the door to tackling a variety of complex questions. Some of the interviewees are interested in the problem of the human and the non-human, the natural and the unnatural while others investigate the grey area between the living and the nonliving or the macroscopic and the microscopic. Is it possible to work and to communicate in this grey area in a more malleable and fruitful way by integrating arts and sciences? Can this help answer complex questions such as the differences between living structures and inert materials or whether life can be imitated by a device? What is the "natural" state of things, from cultural and scientific points of view? What constitutes ecosystem thinking? What does this imply for how to re-perceive the world or to interpret its complexity?

The Western idea of a hierarchy amongst living things is questioned by other interviewees. They raise questions about the microbiome,[2] not only as a paradigm shift within medicine, but as a way to conceive our identity and proximity to others. Today, the concept of the microbiome undermines strictly deterministic interpretations of DNA and genomics. Some wonder about how to recognize ways in which something may be alive in other contexts, such as in earth's primitive systems or on other planets. These art-scientists are aware of the tensions and potential consequences of new technologies and scientific programmes and how the environment is being modified by humans. How can an environment enriched by technological devices and gadgets, become part of the body? Are we accelerators of certain natural processes that will occur regardless of whether we interfere with them or not? Is nature an obsolete technology?

Another issue of interest concerns the relationship between the mind and the body, with its complex, multiple edges. Some interviewees seek to find connections between the tangible and the intangible and tackle the question of whether as a mind, the body can also "think". Or they question the cultural and biological determinants of our behaviour. Others inquire into daydreaming about particles, electrons or

[2] Following (Rogers 2023), the human microbiome is "the full array of microorganisms … that live on and in humans"… The genomes that constitute the human microbiome represent a remarkably diverse array of microorganisms that includes bacteria, archaea (primitive single-celled organisms), fungi, and even some protozoans and nonliving viruses. Bacteria are by far the most numerous members of the human microbiome: the bacterial population alone is estimated at between 75 and 200 trillion individual organisms, while the entire human body consists of about 50 trillion to 100 trillion somatic (body) cells. The sheer microbial abundance suggests that the human body is in fact a "supraorganism," a collection of human and microbial cells and genes and thus a blend of human and microbial traits." Rogers K (2023, June 7, par1). Human microbiome. *Encyclopedia Britannica*. https://www.britannica.com/science/human-microbiome

energy, and more generally about that which is still ignored or only gradually becoming understood.

Other art-scientists delve into the biases of science understanded as a cultural practice, particularly on issues like gender, and on how to better understand the political elements involved in scientific studies. Questions also arise about how cultures, ideas and languages are shaped and how models of ethical responsibility might be built and undone. Do ethical perspectives advance at the pace of biotechnological inventions? Do we have a cultural framework to criticise or to react to these?

Even though these motivations are presented separately, they relate to a common will to broaden personal and public learning from the results and the process involved in the work.

5.2 Scientific Topics and Lines of Research

The specific areas of knowledge used in the selected works and the process of creating them is another category used to analyse the discourse of the interviewees. The art-scientists do not always identify with only one discipline, often they are involved in several.

Some interviewees strain the limits or boundaries of their disciplinary knowledge through their artworks, questioning the idea that disciplines reach objective and hermetic truth in relation to other types of knowledge that may intersect with those. Biology is the discipline with the highest representation amongst the artworks, considering its various specialisations: neurobiology, biochemistry, ecology, astrobiology, primatology, human microbiome and cytomorphology, amongst others. Physics, mathematics, chemistry, artificial intelligence, and experimental psychology also appear.

Amongst the selected works, many combine different disciplines and, as we will see below, in a wide range of ways that produces fascinating results. **Gilberto Esparza** combined several scientific disciplines in his work, focusing on the adaptive processes of bacterial communities in the context of the metabolization of wastewater. Esparza collaborates with the Institute of Ecology of the Universidad Nacional Autónoma de México (UNAM). His *Nomad Plants* (2008–2013) presents a tension in its execution: it uses a very expensive robot which, despite its cost, is not designed to "solve" the problem of wastewater, given its scale. Rather, it is a functional model for the generation of knowledge and an artistic interpellation based on the concerns and interests of the author.

Likewise, **Juan M. Castro** works in chemistry, astrochemistry and microbiology focusing on supramolecular aggregates and soft or fluid materials. He studies the active behaviour of these supramolecular aggregates and the chemistry of materials that manifest organised and dynamic behaviours, that is, those that move, self-divide or have a kind of proto-metabolism. Castro is interested in the idea of abiogenesis, a process that occurs when matter and its molecular interactions shape initial organised systems.

The interest of some interviewees regarding issues associated with the microbiome is clear. Scientist **François-Joseph Lapointe**, a biologist and artist at the Université de Montréal in Québec, Canada, specialises in phylogeny and population genetics, both issues related to evolutionary biology. His doctoral training was in biostatistics and bioinformatics, and he has recently focused on studying and performing with the human microbiome. **Kathy High** is an interdisciplinary artist at the University of Buffalo, USA, who, in collaboration with scientists, works on living systems, e.g. microbiome and immunology, and in the political, social and ethical dilemmas related to biotechnology.

Susana Soares, was trained as a product designer at the College of Art & Design in Caldas da Rainha, Portugal, and currently is Senior lecturer in the Department of Engineering and Design at the London South Bank University in the UK. In her work she has speculated about the genetic modification of plants, the development of new foods based on insect protein and, as presented in Chap. 4, the use of the powerful odour detection capacity of bees to identify diseases. Her reflection on the future of scientific applications opens questions about nature and its mechanisms, and on the ability of biochemical synthesis and other biotechnologies to adapt life to new environmental conditions.

In a more pictorial vein, **David Goodsell**'s illustrations allow him to describe microscopic biological phenomena with dexterity and great thoroughness. Goodsell, a BSc in biology and chemistry and with a PhD in X-ray crystallography of DNA, has dedicated his life to studying structural biology as well as scientific illustration. Thanks to his work and the use of digital techniques, he has been able to describe the evolutionary line of SARS-CoV-2 through sophisticated detailed images.

Špela Petrič, an artist and former PhD researcher in biology at the University of Ljubljana, Slovenia, has used different conceptual, experimental and expressive models, like plants, zooplankton and artificial intelligence algorithms, reflecting on evolution, biotechnology, proto-life or humans life in space.

Dmitry Bulatov, artist, theorist, curator and author of scientific articles, focuses his line of research in both art and science, and uses robotics and biological art as expressive tools. His career in aircraft construction at the National Civil Aviation University of Riga, Latvia (former Soviet Union) allows him to understand the essence of technological systems, while his interests have shifted to the relation between the living and the nonliving.

Andro Montoya's PhD was centred on neuroinflammatory mechanisms, particularly the intimate relationships between the central nervous system and the immune system. For him, this relationship not only involves the potential of new approaches for healing diseases but is also a boost for creativity. In his graffiti he explores the "phylogenies" of the hybrid and speculative creatures that arise, almost spontaneously, at the place of their production.

Nicolas Desprat, is a biophysicist who works with bacterial communities and in-host pathogen interaction. **Jean-Baptiste Boule** is a biologist by training, but very interested in physics. Together they work on the physiology and molecular nature of adaptation in microorganisms, and how physical and chemical constraints (mechanical, temperature, pH) build and rebuild microbiological communities.

Manuel Théry's scientific interest lies in the study of cellular architecture mediated by the network of microtubule filaments that give shape and movement to the cell. In this way the internal organisation of cells is consistent with external conditions, for example, in relation to the geometry of the substrate.

Michail Rybakov mixes artificial intelligence with cartography to generate maps that function as living, complex and dynamic entities, rather than static descriptions of a natural phenomenon. In his works, he resorts to ecological interactions, historical scientific illustration, design and the body.

From the perspective of evolutionary biology, **Perdita Phillips**' proposal starts from her fieldwork in ecology and soil sciences, at a crossroads between the natural world and the artificial world, identifying patterns and searching for correlations between the two.

Artist **Rachel Mayeri** has worked in primatology, comparative psychology, social anthropology and gender studies, amongst other areas. Her interest is focused on achieving a better understanding of human beings through the study of primates.

The world of abstraction, equations, physics and mathematics are places of research for several selected authors. On the other hand, as a mathematician **Jimena Royo-Letelier** is interested in the behaviour and variation of the parameters of an equation. She focuses her research on spectral analysis, machine learning, signal processing and its relationship with animal behaviour as well as neuronal activity. **Daina Taimina,** a mathematician and associate professor at Cornell University, USA, began to give material life to equations by creating crochet fabrics. Her aim is to make hyperbolic geometry visual and tactile. Transposing these planes to a spatial figure allows her to teach hyperbolic geometry in a very practical way.

Yunchul Kim is an interdisciplinary artist who works using colloidal suspensions to understand the physics of continuous media. He is currently interested in the symbol of the "knot" and its meaning in physics, mathematics and the process of DNA replication. **Shanthi Chandrasekar** is interested in cosmology, human anatomy and neuroscience. She initially studied physics, then psychology, attracted by the functioning of the brain. She combines this knowledge with elements of Tamil culture.

The theme of behaviour and the stimuli/response problem is recurrent in the interests of some of the interviewees. Researchers **Suzanne Dikker** and **Suzan Tunca** focus their line of work on cognitive neuroscience and dance interactions. They want to learn how the human brain operates in dynamic, social bodily experiences and use data-based analysis systems in the process.

Nicholas Wade is interested in cognition and visual recognition systems. Over the course of thirty years of research, Wade has drawn on the distinction between what he calls the "naturalistic view" (perception in the natural world) and the "experimental view" (which takes places in controlled laboratory conditions) to study various aspects of visual perception, for example, visual stereoscopy.

Paul Vanouse is a "radical interdisciplinary" artist that uses techno-science as a medium to address complex issues regarding techno-science itself. Focusing in biotechnology, Vanouse explores genomics through molecular biology techniques as DNA sequencing. More recently, his work problematizes human border and identity working with the microbiome.

Finally, **Max Liboiron** focuses on environmental issues, in particular studies on plastic pollution in different ecosystems in Newfoundland and Labrador, Canada. Liboiron is also interested in the development of feminist and anticolonial scientific methods of indigenous origin which are grounded on humility, responsibility, and harmonic relations with the land.

5.3 Materiality and Artistic Methodology

This category refers to the use of different materials, techniques, devices and procedures for the elaboration of an art-science piece. Behind the selection of a particular materiality, the personal motivation of an author together with their scientific disciplinary inclinations was found. It is important to note that the choice of materiality affects the methodology used in the creation, just as the methodology used affects the materiality.

For example, choosing a bioreactor as the materiality of an artwork requires a particular methodology, quite different from that which would be employed if working with pen and paper. This bidirectional influence is itself diverse, and depends on disciplinary factors, experimental lines of research and a variety of instrumental techniques and procedures. This relationship is crucial for its impact on the interdisciplinary creative process of the artistic piece.

This diversity goes from installation, to computer graphics, from film to Public space intervention. A performance is devised with humour and speculative thinking, testing new experimental conditions. An experiment takes place in the middle of a museum, expanding (and restricting) the technical scope of a laboratory *set-up*. Microscopic and living entities (cells) are coupled with macroscopic nonliving objects (a building). Motion-capturing technologies are integrated with the movements of a dancer; shared taxonomies are developed between dancers and scientists; an algorithm related to the synchrony of dance movement is conceived to reflect simultaneity.

There is also music and noise. Data is sonified. Fused with musical pieces. Ground signals are musicalized and even the cross section of a geological map is converted into sound. Multi-screen video installations are mounted, and experimental films and documentaries are made. Photography takes part with high contrast large format images.

Biology finds its way in this journey of methodological and material subversions: geometry is made with the shape of cells; programming codes are applied to living matter and metabolic metaphors. Works involve saliva, blood, semen, milk, organs, bacteria, chromosomes and DNA as well as aquatic plants, self-photosynthetic systems and contaminated water. Hybrid systems between machines and organisms are developed. The gut microbiome is used and observed, the bacterial expression of faecal microbes are transplanted. Scientific protocols may be followed in an anarchical way involving, at the same time, the scientific collaborator as part of the materialisation of the work. Inspiration is taken from methods of design, the social sciences, behavioural sciences, and ethical dilemmas. New methods attract new people.

Networks are created between artists and scientists that last over time, seeking new concepts and expressive strategies. Some search for the poetry underlying scientists' research work. In the field work, similarities and differences between the contemplative and the analytical standpoint are contrasted with respect to the direct encounters with nature.

Scientific information is rethought to communicate it, for example, by bringing together feminist ideas and anti-colonial science. Charismatic representations of scientific objects are executed; traditional materials are joined with cutting-edge science. Results are resolved by industrial or graphic design.

Intuition is followed; data is visualised but without the usual precision. Public feedback is heard and considered, working from a new ethics with a material base; scientific truths are submitted to speculation. Topics are investigated by deeply reviewing the scientific literature and field observations, including open interviews.

Some of the art-scientists start by writing. Others speculate on the wild. They approach an individual as a plant in their multiple dimensions: molecular, cellular, personal and social. Or use materials with dynamic spectral patterns linking the artwork to aesthetics or philosophical anthropology. They offer workshops and make interactive web pages, create speculative works linked to everyday life collaborating with companies or small developers to face technical challenges. They look for common definitions and coin new ones.

5.4 Differences Between Artistic and Scientific Approaches

When scientists and artists come together on a project, their differences and contrasts initiate and maintain the interdisciplinary discussion. In the case of the selected works, all the creators knew something of both art and science. Some of the scientists stopped practising science and declared themselves artists, while others carry on with their professional scientific activities in parallel with their artistic works. Those with formal training in art have developed great skill and expertise in one or more scientific practices.

For Špela Petrič, artistic practice is concerned with cultural construction, and thereby with identifying underlying assumptions, their consequences and ways to act accordingly. On the other hand, she thinks the sciences have powerful narratives that shape time and space. It is in this way that Petrič sees herself as an artistic researcher, seeking experiments conducive to rethinking possible futures. In a similar line, for Michail Rybakov the correct representation of data is a must in scientific practice whilst artistic representation allows for freedom to create a visual effect that can be more powerful than scientifically “correct” representation.

Esparza notes that throughout history art has created knowledge and a record of the cultural, political and social processes that have taken place in different contexts. For him, something similar is true in science, if we consider the expanded and still expanding conceptualization and practice of the scientific method. That is, when the scientific method integrates other approaches, for example, art, philosophy, or indigenous systems of knowledge, the possibilities for understanding the observed phenomena are valuably broadened.

Considering understanding in term of references to previous achievement, Lapointe posits that whereas in science an article published a couple of years ago is a legitimate starting point (no need to go back to Darwin, so to speak), in the humanities the original source is often used to deconstruct or critically review an idea. If science becomes confusing, it is rejected, but art can afford to be fuzzy.

For Petrič, in artistic practice a multiplicity of connections are created, things become more opaque and there is an abundance, sometimes even an excess, of possible meanings. Nicholas Wade identifies a distance between art and science insofar as science simplifies the object of study so as to manipulate one variable at a time. The nature of the artist's enterprise, on the other hand, goes in the opposite direction: it complicates, intrigues and can even baffle the observer. In methodological terms, Rachel Mayeri recognizes that although in both art and science the procedures are experimental, scientists delve more in the method to answer very precise questions.

Dmitry Bulatov has a similar perspective. He claims that in the arts it is unknown how exactly a complex system will react or behave, so the art-scientist takes a first step without knowing how the system will respond. Art employs aesthetics and form in experimentation and puts them in relation to both living organisms and the public. Methodologies in art are difficult to understand and are continually reinvented, so you often hear the question "but is this art?", unlike what occurs with science, where the scientific method is clearer, and it is easier to know whether or not something is science. For Mayeri, science is about outlining facts and creating knowledge within a certain community, with conventions about how to produce data, write articles and publish books. Connection with art produces an openness to the meaning of scientific work and permits a release of "repressed" motivations of scientists interested in art: scientists are often weird and curious, and some have a high degree of imagination (as do many artists).

From the field of design, one of the differences between art and science that Susana Soares identifies is that science seeks to formulate a hypothesis and see to what extent it can be proven or falsified. Design, on the other hand, seeks to create an object that people can use, studying the factors of human interaction with the world. Bulatov expands on a similar idea when noticing that although in art there are no hypotheses, art presents a thesis and an antithesis which are offered to the viewer to choose. Art creates an unstable system and the art-science work offers variability and defines limits to science. Kathy High proposes that the artistic process does not attempt to verify a hypothesis but rather is more like a learning experience, an active process that is gradually understood through the production process itself. In terms of exhibition, High is interested in exposing her experimental system so that the viewer has a first-hand experience of the subject matter, which in science is usually invisible to non-specialists.

For Royo-Letelier, science attempts to establish a truth, prove a theorem or a proposition by experimental or statistical means. Artistic production, on the other hand, can be interpreted from many points of view, which for her is wonderful. Shanthi Chandrasekar claims that for her, art is complex because it devises questions that probably cannot be answered. For Yunchul Kim, the process of artistic research is quite intuitive and is directly connected to the senses of the body, as well as being affective and performative.

For Perdita Phillips, the world is usually more complex than hypothesising, collecting data and analysing it. This complexity appears from the moment a proposal for a research project is written: a lot of science that is structured and executed bearing in mind funding and finance. From the perspective of contemporary art, financial support is always a problem. Bulatov comments on a mechanism whereby people partially own an artwork through shareholding. This is unthinkable in science. Petrič remarks on how the possibilities of financing collaborations in science far exceed those in the arts. The sources and mechanisms of funding art and science are drastically different.

Another issue interviewees discuss is the maintenance and conservation of the piece of work. Unlike science, whose final product is an article, most art pieces are unique, must be preserved over time and be exhibited in different places. Bulatov comments on how owners of art-science works are required to keep them "alive", in literal terms. For Petrič keeping living things on display is a performative action where everything must operate correctly, so that it works as expected. However, this almost never happens as many conditions are constantly changing: the specific environment of the exhibition space, the duration of events, the type of maintenance required, the season of the year, the influence of the public. Under these circumstances, experiments, living things and even technologies can escape control. In her view, there is a story, a narrative of the work, that will never behave exactly as one wants.

From the point of view of experimentation, there are big differences between art and science. Accidents, chance and uncontrolled situations are part of the artistic process. According to Bulatov, while scientists are uncomfortable when the experimental object "makes" mistakes or behaves differently than expected, for the artist this is a resource. He claims that one reason he left his job in aircraft construction was that a single error could be fatal. Art inhabits a conceptual limit and the reality art model is always off limits to what appears to be objective and rational. There are differences between art and design too. While design confirms our realities when it creates friendly shapes or spaces, art attempts to discover and reach the limits of design. In this way, art can function as methodological criticism in different fields. Suzanne Dikker and Suzan Tunca work from science since parameters and coordinates define their practice in a conscious way. But they use art because it operates on the basis of other taxonomies, such as intention or synchrony.

David Goodsell contends that art can evoke an emotion in the audience even regarding science itself. For François-Joseph Lapointe, when you are doing science, the aim is to successfully convey understanding to colleagues and peers, not the general public. On the contrary, in artistic practice the objective is for the audiences to question and challenge themselves. It is accepted that art can sometimes inspire feelings of insecurity or even emotional instability in the audience.

According to Jean-Baptiste Boulé, in psychological terms artistic practice can relieve tension and provide space for freedom of thought. In his mind, scientific work is more stressful than art, as scientists fight against things from within as they try to keep things under control, constantly wary of being invaded by intuition. For several of the interviewees, although they do not know how it happens, art somehow conveys emotions when you look at something and feel that it is beautiful or upsetting. On the contrary, when you do science, you want and need to work with minimal emotion.

For Wade, science works within a series of rules that must be observed if peers are to be persuaded that results are replicable. The artist need not care about replicability. The artist is concerned with novelty. In terms of representation, if you do science, you make simple or figurative images. When art is made, complex images are elaborated. Wade draws a distinction regarding the rules for achieving success in art and science. For Bulatov, scientific art and technological art in general, need not be considered from a perspective of progress, usually referred to as the main criterion of success for scientific practice.

The freedom of art allows the generation of open analogies of science with different fields. In the example provided by Chandresekhar regarding Indian philosophy, there is a goddess of energy. She states that if you translate a few phrases of Sanskrit verse and show them to a physicist, it's possible they will conclude these refer to dark, potential or kinetic energy.

Boulé and Desprat claim that science is devoid of political motivation while Liboiron remarks that there are many artists who touch upon political themes who nonetheless are not necessarily ethical. There are artists who kill animals to defend a position about animal rights. Or artists who work on slavery and nonetheless reproduce sexism when they are doing anti-racist work. Artists do not ask people for permission or consent. At least in science, Liboiron contends, there is a code of ethics. Although not necessarily perfect, at least it exists.

As a programmer, Rybakov can work both for industry and for his own artistic purposes. Many of the programming codes that function commercially find their way into his disruptive works of art. Paul Vanouse comments that when two worlds come together, there are ways to articulate them using a philosophical discourse to foster social change. The theorization and the object come together in a single project that allows one to start working with the art piece.

In turn, Phillips delves into the academic context of science and art. While artistic experimentation is carried out in the studio and consists of putting together different things and seeing what the results are, data obtained through scientific experimentation usually has a very limited range of interpretations. The results of artistic experimentation have a much broader potential for multiple interpretations that can be encompassed in a creative production. This more speculative approach to data is much more difficult in science. Phillips contends that when working in academia, it can be quite restrictive for art students to be presented with closed questions, but at the same time it is useful that they focus on delimited projects.

Taimina notes that when someone external to the art world enters it, especially if they come equipped with an academic background, artists sometimes feel "insulted". At the same time, she mentions the personal experience of entering the scientific community with her art-science proposal and being met with a "that is not real mathematics". In other words, the art-science interaction in academia can generate points of friction between the most conservative positions in both worlds.

From the point of view of the assembly process of the artwork, Boulé and Desprat emphasise its intensity, specifically spending up to a week totally immersed in the enterprise. Scientists have restrictions that artistic spaces do not necessarily contemplate and therefore it must be accepted that the work might not be exactly what they had intended it to be. Goodsell comments that when you assemble an artistic work, things are infinitely more difficult than what he is accustomed to in the laboratory, and a lot is invented "on the fly". Similarly, Leiboron comments that artists believe they can do anything, however, often technical issues become fundamental to resolve.

From the perspective of communicating science, Juan M. Castro wants to see more real-time demonstrations of findings and results, not only published in papers but presented in front of live audiences, as was done in ancient times. There are many publications that do not specify the intricate details of experimental processes nor, for example, if the procedure functioned only on one occasion. In his opinion, given the tremendous amount of work done in laboratories, it is a waste of time and energy for it to remain only within the bubble of science. The public should have access to the material and performative knowledge.

For Petrič too, art-science works must not only be exhibited in public spaces, but the experiments themselves should also take place in the real world. For her, when working in the field of art, experiments are not really complete until presented to the public. Only then is it possible to know their success or failure. This is also the most rewarding part of the work since reception is ultimately unpredictable and continues to unfold with new and different audiences. It is this excitement that motivates her to continue doing art rather than science alone. Moreover, in her opinion art allows for a quicker, more immediate response to changes in society than does science and it reaches a far greater number of people.

Montoya refers to another element that distinguishes arts and sciences: the satisfaction of the aesthetic experience, which manifests itself more directly and intuitively than the intellectual satisfaction of science. For Yunchul Kim, the experience and manifestation of aesthetic events relates to our minds, bodies and feelings, which are not involved in the appreciation of science. Works of art fluctuate on the horizon of our mind beyond the knowledge and horizon of science.

5.5 Art-Science Synthesis: Its Meaning and Implications

Each of the art-scientists considered in this book has a particular way of conceiving the link between art with science. Identifying these may help to better understand and value the contribution of art-science synthesis as a source of knowledge and critical practice. Remarkably, in each work of synthesis, novel questions, processes and methodologies are involved and developed.

The production of art-science works contains or reveals the biographical, social and political subjectivity of its creators, together with their academic background. Works often resort to fiction and speculation as well as to demonstrated facts, experiments and personal experiences. They embark on processes within which human relationships, learning, informal collaboration, and translations or transpositions from one domain's practice to another are important components. Desprat observes that it is easier to step from science to art than it is from art to science but less likely to find scientists who are interested in art than artists interested in science. For Jean Baptiste Boulé, sometimes scientific forms give rise to the best works of art. That said, it is not easy for an artist to get involved in science just by reading books or articles of a scientific nature. It is key to work and collaborate with scientists, ideally outside the institutions of science. For Phillips, key to really getting to know scientists directly and informally, are networks of science "aficionados" where dedicated amateurs and professional scientists meet outside of academia. In the art-science process, the hidden philosopher may appear within the scientist.

For Kathy High, working seriously and systematically in biology labs was both rewarding and productive. There are similarities between art and science labs, she claims, such as interest in the visual, form, complexity and the living and nonliving. She thinks the artist adds layers of cultural content to the biological experiment and by adding nuances to the phenomenon investigated turns it into a lived experience.

The integration of art and science articulates and exposes in a complex way not only materiality, but also dimensions of the self, emotions, collaboration, curiosity, compassion, knowledge and subversion, amongst many others. Together these elements intertwine and trigger different, diverse meanings. The materiality of a piece, for example, stimulates emotions about the scientific motif it represents and, in this way, science is poetized or metaphorized. Or new strategies are explored to extend the laboratory into the exhibition space. Experimental results are delivered in different perceptual and poetic modalities that make palpable the emotional basis which justifies the work from both the scientific and artistic viewpoints.

Art-science integrative work makes it possible to represent models, theories, conceptual and mathematical objects from new platforms, for example, using sound, movement or installation arrangements. For Royo-Letelier, art-science practices allow us to expand the aesthetic feeling of an equation to new patterns of natural or abstract referents. She attempts to share with the audience the pleasure of the researcher when exploring nature or mathematics and to stimulate scientific intuition. Just as equations can be simple, beautiful and elegant, a sophisticated drawing can lead to deep thinking. In a sense, this is Chandrasekar's definition of her aesthetic: her shapes denote visualisation regarding abstract concepts, theories and objects that are difficult to comprehend, observe and prove, like infinity, black holes and dark matter, and reveal her extensive knowledge of physics. She considers these shapes beautiful, but they also contain much scientific information. Chandrasekar works on a plane that is equally influenced by art and science without belonging to one or the other. For Yunchul Kim there are key transdisciplinary and interdisciplinary concepts that cannot be neatly separated into science, arts or humanities.

Lapointe calls his synthesis "para-disciplinary" because it employs both art and science in a mutually "parasitic" way, in a parallel and synchronous process. For him, the key is symmetry and non-instrumentalization of either the artist or the scientist so that both feel that they contribute equally to the endeavour by teaching, learning and attaining mutual understanding. Soares calls collaboration "co-production" and, amongst other aspects, emphasises that the degree of speculation involved in the final piece of work is negotiated between artists and scientists.

At first, Lapointe's scientific work was the inspiration for his art, but as time went on his art began to inspire and fuel his scientific work. He employs the same data to produce both the artistic work and the scientific publication and often these joint results inspire the science of others, as well as both philosophical and artistic interpretations. Théry comments that the artistic performance or staging of his work is an experiment in itself, and as such avoids the repetitive cycle typical of artistic installations. In his experimental performances, previously untested conditions are applied, which can lead to discoveries or to a reorientation of lines of research. For him, if science is not undertaken during the artistic piece, the process is meaningless. Likewise, High explains how a renowned scientist used the research they had done together in his own research, and that their collaboration had helped him formulate new questions.

When artists enter science, they help to loosen, open and alter what are initially more restricted scientific questions. Artists have more freedom to take speculative leaps, to articulate meanings about the scientific object, to put variables together in different, novel ways. The artistic operation uses experimental data as an input for other sensory spaces. To inform and raise awareness amongst a wider audience about emerging technologies and scientific advances, Soares combines biology and technology to explore new critical and speculative functions for design.

The traditional bubble of the closed circulation of works of art on the one hand, and of scientific publication on the other, is broken. Artistic performance becomes a pretext to test new tools, as seriously as they would be employed in the laboratory, removing some restrictions while adding others (for example, staging a production for a particular audience, or creating a certain dramatic appeal in the staging). For Castro, art-science work must have aesthetic magnetism and function according to the objectives of research. In this way, art-science broadens the horizon of problems, for example, to include an ethical angle, which can produce a powerful impact on the audience.

For Théry, when the scientist who makes art does not have an artistic background or training, they enter a dilemma between humility, arrogance and confidence. Despite not knowing exactly what they are doing from an artistic point of view, the scientist can work conceptually, questioning and expanding the limits of a scientific experiment. The scientist has opportunities to open conversations between disciplines and with the public, without resorting to previous references or experience, and can do so with self-confidence and without shame or arrogance. This implies the personal challenge of taking a naive, inexperienced, distanced (yet valid) approach to the scientific problem, as might a child, for example. Art-science recovers innocence in art and science. It allows the scientist to evolve, to change perspectives. It allows artists and scientists, as well the public, to enter otherwise closed worlds.

Wade was led by copyright restrictions to create his own images. He uses artistic tricks, for example, figurative vagueness, to generate the feeling of "visual achievement" in the viewer, at the service of making the motif or concept easier to remember. He works the history of stereoscopy stereoscopically. He studies visual phenomena, such as binocularity, making images in an artistic way but crucially, he is uninterested in whether his work is classified as art or science.

It is hard to claim that a scientist who makes art is an artist as it is to say that an artist who does science is a scientist. For Desprat, what you produce or create determines whether you are a scientist or an artist. Royo-Letelier contends that unlike artists, the scientists who integrate art and science need not justify themselves as artists, except for very specific purposes, for example, for gaining external recognition or financial support. In hybrid art-science work, distinguishing artistic value from scientific operation is not necessarily nor usually the point of view of the creators.

Art-science work must be robust enough to spark the interest of diverse audiences and provide them with opportunities to express their ideas and opinions. For Royo-Letelier, art-science symbiosis requires solving new and complex problems by both understanding the specialised literature and through the process of creating the work itself. In this sense, Paul Vanouse must keep up with the ever-changing tools of the biotechnological landscape and the ethical challenges they pose. Only in this way can he help others with their work, particularly when the focus of criticism is both the subject and the artistic medium (for example in bioart).

According to Taimina, whose artwork is defined by hyperbolic equations, the piece and the scientific concept behind it are one. Théry warns that in art-science integration, a pretty image of biology, for example, is not enough. The challenge for art-science is to be as properly artistic as it is scientific; works must have an intention, raise open questions, speculate and seek change. That is the real value of art-science, regardless of whether it is beautiful or not, in his opinion.

For Dikker and Tunca incorporating complex, difficult concepts in artwork knowing that audiences lack any specialised background is a central challenge. Scientific taxonomies need to be transferred into the artistic work. Likewise, a piece may contain terms with problematic connotations for the scientist. They must reduce notions, conversations and debates to an aesthetic experience that can be accessed without the need for specialist knowledge. Sometimes the terms mean different things to different people in the audience. Original, inspiring questions from the audience appear and in Dikker and Tunca's view, it is worthwhile to take them seriously.

Bulatov argues that art-science practice is a third way of thinking and working, uncomfortable for both art and science. It questions the meaning of the term "progress" and society's control over technology. This hybrid approach seeks to recast myths and archetypes, and the subject-object relationship. Petrič wonders whether it is possible to use scientific methodology to de-objectivize or subjectivize the organisms under study. She also wonders if scientific materialism has reached a limit as to what it can teach us about the world and our relationship to it.

The synthesis of art and science is supported by academic disciplines such as Science, Technology and Society (STS), Media Archaeology, and Cultural Studies. And coherent with scientific pluralism, the "war of the sciences" of the 1990s, and posthumanism. Vanouse adds that the intellectual basis of the art-science endeavour allows to distinguish the critical positions of these pieces from the current "anti-science" movements, (for example, QAnon in the USA). According to Lapointe, the integration of art and science allows us to approach many themes from multiple angles: magical, biological, mystical, spiritual, all of which converge in the piece. Michail Rybakov proposes that art is a way of thinking, that it can be applied to different activities from science to industry and that the resulting work is a way of communicating that idea. Following this line, Rachel Mayeri believes that science-inspired art can exploit the passion scientists feel for their subjects in ways they cannot explore professionally. Artists can bring out the cultural meanings of science and its problems. They can raise questions and discussions, for example about ethics, in a non-confrontational or accusatory way.

6 Action Points and Perspectives

Abstract

This chapter invites the reader to get involved in an art-science project. Some easy to follow steps are presented as a guide to create, from their own scientific research, an art-science work proposal. Several art-science current initiatives are also presented with a list of salient art-science graduate programs around the world.

Keywords

Integrated curriculum · Interdisciplinary approach · Art-science project

This book is an invitation to bridge the arts and the sciences recognizing some of the challenges they face today. In navigating this complex terrain, some categories were developed that illuminate dimensions of the art-science integration. With revealing examples of current art-scientists, the dynamic interplay between arts and sciences not only nurtures and enriches both fields, but also engenders a plethora of thought-provoking inquiries across diverse domains.

This final chapter serves as a practical resource for scientists who want to embark in an art-science exploration. The deep involvement and the theoretical, conceptual and technical knowledge of scientists in their research area, have more than enough potential to deliver a powerful narrative through an art piece. In this sense, the conceptual spaces opened by contemporary art are a fertile field for both complex or spontaneous reflection and to express the personal excitement involved in the scientific practice.

The authors have had the privilege to guide senior scientists and graduate students through art-science workshops inside the lab, intended for the creation, proposal and defence of art-science projects. They learned that such a project can be developed by any scientist with the spirit and ambition to take risks regarding subjective involvement. Or to speculate and expand the scope of reflection and dialogue associated with their research area. These types of projects may become an excellent platform for communicating and circulating their research in a unique and engaging way, opened for new expositive spaces and audiences.

The examination of contemporary art expressions, taking a distanced position to the research space, and, as discussed in Chap. 2, exploring the metaphorical expressions involved in each scientific topic, uncovers novel connections and symbolic meanings, contributing to a more profound or divergent understanding of research.

At the end of this chapter, a comprehensive list of initiatives developed by institutions or independent groups is presented, where the ongoing debate of art-science integration is unfolding. As well, a series of academic

© The Author(s), under exclusive license to Springer Nature Switzerland AG 2024
M. Velasco and I. Nieto, *The Art-Science Symbiosis*, https://doi.org/10.1007/978-3-031-47404-0_6

programs for graduate studies in this rising field is also presented.

6.1 Work on Your Own Art-Science Project

If you are a scientist and you want to put your hands and soul in your own art-science project, you may follow the next steps:

1. Find a group of colleagues (from two to five) that want to start an art-science journey from their own scientific work. It is not easy to find adventurous fellows, but working alone in a novel area may be difficult for some people.
2. Become more familiar with the history and manifestations of contemporary art, exploring interdisciplinary artworks in particular. The bibliography presented in this book could be a starting point.
3. Assess your current experimental system (the lab space, equipment and machines, instruments etc.), observe it from a distance, with a defamiliarized view. Ask yourself about your subjective feeling or taste to these objects: Do you like/dislike them? Do you feel comfortable and proficient with them? Do you love its shape, its smell, its colours? Extend this view beyond the lab, to the building, your colleagues, institutional authorities, etc. Write them down frankly as an automatic writing process or as a free essay.
4. Make more automatic writings (with restricted time, let's say, three minutes per concept) relating your research work creatively with powerful concepts, for example: money, romantic love or global crisis.
5. Find a collaborator from the arts. Try to delve into the passions, pains and dreams of this collaborator, and share your own with them.
6. Identify and write down the metaphors used in formal or daily life discussions about your research.
7. Once you have identified the most relevant, write down opposite metaphors (e.g., if the usual metaphor denotes war, apply collaboration; if it denotes passiveness, use activeness; rigidity or stability, think about plasticity or change, etc.). Try to exaggerate the usual metaphors to the absurd, force eccentric metaphors and see what ideas or imageries may appear in relation to your research. Speculate on possible re-interpretations of your results based on the new concepts raised.
8. Critically review the arguments, personal and public, that motivate, justify and validate the research program you are involved in. Related to these findings, write down possible futures derived from your research. Speculate about what would constitute an ideal experiment.
9. Search for digital or material images that visually represent the metaphor, imageries, speculations this way raised, and compose them in a digital or physical collage. Use any image or video you may find, on the internet or other sources (magazines, prints, graphic books, newspapers) to support your proposal. This exercise could become your first artistic piece.
10. Now explore and compose a more complete work: a scientific poster, a logo, a short video, descriptive sketches, etc. Look for references and inspirations in art books or the internet. Ask an artist for help. Use the media that is more comfortable or attractive for you. Try more than one. They may become part of your artistic portfolio.
11. With all this material and training you will have enough information and resources to propose an expository piece. This proposal should be ideally publicly presented to a broad audience. In this way, you will expose yourself in an expressive/declarative exercise and be forced to defend your project in an open dialog.
12. Get in touch with us to share your proposals and visit our website: www.artscience-symbiosis.net

velasco1340@gmail.com
ignacio.nieto@ug.uchile.cl

For further study on art-science integration see,

- Cohen R S, Renn J, Gavroglu k, (ed) (2010) *Beyond mimesis and convention.* Springer, Dordrecht.
- Reichle I (2008) *Art in the Age of technoscience.* Springer, Vienna.
- Born G & Barry A (2010) ART-SCIENCE, *Journal of Cultural Economy*, 3:1, 103–119.
- Ede S (2005) *Art and Science.* I.B.Tauris & Co Ltd.
- Miller A I (2014) *Colliding Worlds* W. W. Norton & Company.
- w/k - Zwischen Wissenschaft & Kunst Press. A peer-reviewed online journal
- Art and Science. ISTE Open Science Journals (in French).

6.2 Current Art-Science Initiatives

This section presents a comprehensive review of current initiatives in the art-science field, useful to learn more about art-science projects. This exercise offers a panoramic view on the diverse focuses these current initiatives may have.

Latin America

In Chile there have been several instances where to integrate art and science. The **Bienal de Artes Mediales**, focused on new media artworks. There are institutional an bi-national agreements too. For example, since 2018, the agreement signed by Ars Electronica and the Government of Chile to send artists and works. The National Funds for the Arts (FONDART) finance Science/Art projects. In 2017 the first **Art Science Symposia** was held in 2017 at the **University of Atacama**—Chile, led by the art-scientist Ph.D. Alejandra Tello, who works on nanoscopic and microscopic material for art. Later at the **University Los Andes**, Santiago-Chile, developed a second Art-science Symposia, led by the Immunologist PhD Karina Pino.

From an ecological perspective, there is the **Studio Vegetalista School** in Chile, which focuses on experimental ethnobotanical knowledge of healing with a special focus in projects related to science. A shamanic vision is linked to local rites, where instances of collaborative work are developed at a national and international level. The residency **Bosque Pehuen: Multidisciplinary Research Station for Nature**, by Fundación Mar Adentro, supports artistic projects emphasising the conservation and protection of nature. The **Ring project from Chile BioGeoArte** uses the referents of scientific work, geography and territory, in an interdisciplinary way. The **Fungi Kingdom Festival** led by artist Miguel Bolt, carries out several artistic, cultural and field activities in relation to fungi ecosystems conservation in South Chile. Finally, the first **Bienal Arte & Ciencia 2022**, in the city of Concepción and the joint work with the University of Concepción, U niversity of Bío-Bío and University Católica de la Santísima Concepción, gathers national and international artists and scientists around coastal communities.

In Argentina, these spaces are beginning to be articulated. The **MUNTREF Art and Nature**, dependent on the University Tres de Febrero, promotes artistic projects developed from the academic field. The Cultural Center of Science will carry out the **Bios Project**, where artists and scientists show the state of the art in the field of biotechnology and Bioart. The **Center for Art and Science of the School Arts and Heritage,** of the University of San Martin, conducts research, seminars and articles with multiple focuses, incorporating different perspectives in the interpretation of data, questioning current paradigms. The **Latin American Bioart Lab (LatBioLab)** at the Universidad Abierta Interamericana, in Mendoza, provides laboratory space for collaboration between artists, biologists and engineers. The **GenBA** project is a network of institutions around Art, Science and Technology.

In Colombia, a collaborative network with Europe, called **Mutante.Lab,** organise the

Suratómica Festival. The **FLORA ars+natura** is a space for contemporary art with an emphasis on the relationship between art, nature and the body that, through its Secretariat of the Culture Recreation and Sports, emphasises the Strategic Line of Art, Science and Technology, financing projects that go along that line.

Brazil has developed a focus on locally situated techno-scientific knowledge. **The Green Lab** in the Amazon, is a laboratory that tries to answer how culture can develop new strategies for the expansion of environmental justice in the context of the Amazon. **The Institute of Biosciences** from the University of São Paulo, invokes Nature as a bridge within art and science, with an integrative approach to the structure and evolution of the Amazonian biota. The institute has mounted the exhibition ***Amazônia: Os new travellers***, that draws a parallel between the explorers of the nineteenth century and current travellers in the Brazilian Amazon. Finally mention the space called **Espaço Ciência Viva** in Rio de Janeiro, founded in 1982.

In Mexico the **Call for Art, Science and Technologies program** was jointly launched by the Ministry of Culture and the UNAM. The **Laboratorio de Arte Alameda**, is an exhibition centre in Mexico City that functions in a church building. The **Art+Science group** is housed in the Faculty of Philosophy and Letters of the National University of Mexico.

The **Andean High Technology (ATA)** stands out, contributing to the development of a new culture based on the use and expansion of electronic media in Peru and Latin America. Other example is the **Chair in Design, Art and Science,** an interdisciplinary commitment between Faculties of Art, Design, Natural Sciences and Engineering to dialogue, reflect and search for new lines of research with universities in Chile, Argentina and Brazil, with the perspective to be part of UNESCO Chair.

Europe

Since the early 2000s, researcher Jill Scott and the University of Zurich have developed the **Artists in Lab Project**, consisting of a series of artist residencies in scientific research environments. The dual objective of these residencies was to expose the artists to scientific knowledge and to open up science to potential collaborations with art (Scott 2006). In *Artists in Lab*, laboratories from areas such as biology, computing, engineering and physics were involved. Resident artists reported their experience in the form of ethnographic narrative or social field research. Scott is interested in the role of artists linking science with the public, in the possibility of addressing ethical and social aspects in research and in how to transfer these aspects in new technological devices. Another motivation is the creation of university degrees that would allow more symmetrical relationships between artists and academia.

Since 2014, **Biofaction** is a science communication project based in Vienna. *Biofaction* works on a wide range of projects, from a film festival to the production of scientific articles, from research projects to purely artistic residencies. One of these projects was the residency for artists called *SYNPEPTIDE*, focused on designing and testing new antibiotics produced by a wide variety of organisms.

One of the most renowned cases is the art-science residence system is the **Art at CERN**, of the Conseil Européen pour la Recherche Nucléaire in Genoa, Switzerland, currently the largest particle physics centre in the world. Its motto, accelerating science. Art at CERN is directed by the curator Mónica Bello and has a residency system focused on particle physics. The chosen artists collaborate creatively with scientists and engineers to produce artworks or exhibitions in museums around the world. For example, the Quantum/Broken Symmetries exhibition carried out by Mónica Bello and José-Carlos Mariátegui, open from 2018 to 2021 at FACT (Liverpool, UK), CCCB (Barcelona, Spain), iMAL (Brussels, Belgium), Le Lieu Unique (Nantes, France), National Taiwan Museum of Fine Arts (Taichung, Taiwan) and Kumu Art Museum (Tallinn, Estonia).

The United Kingdom has its own state funding system for the link between art and science. Such was the case of the **Wellcome**

Trust's Sciart program.[1] Running from 1996 to 2006, it was funded with nearly £3 million. On the other hand, in the **Science and the Arts Section** belonging to the British Association of Scientists, different organisations are invited to support arts awards, engaging the public with biomedical science through the arts. **Arts Catalyst,** based in Sheffield, is a contemporary arts organisation at the intersections of art, science, technology and culture, promoting new ideas, dialog and transformative experiences. It differs from other initiatives due to its critical position on techno-scientific development, similar to what the net.art movement or the artistic avant-gardes did in the past. Research of an artistic-scientific nature involves dissimilar disciplines, from nuclear energy to ecology.

SpaceR2 is an interdisciplinary studio for sustainable creative practice located in the coastal town of Ramsgate, United Kingdom. Individual and collective works of art are produced, including research, experimentation, making and learning. **CRASSH** (Center for Research in the Arts, Social Science and Humanities) at the University of Cambridge, is a group that focuses on the social sciences, psychology and philosophy, discussing interdisciplinary intersections between these fields. The recent **Cavendish Arts Science programme**, is a laboratory inheriting the tradition of Cavendish Laboratory created in 1874 to understand the physical world, from electromagnetism, to the division of the nucleus and X-ray crystallography.

Other art- science projects is the contest carried out by the MU Gallery, in Eindhoven, Netherlands. The project is called the **Bio Art and Design Award (BAD)**. This annual contest provides advice on art projects that involve scientific knowledge and design where selected artists work with a laboratory or institution that provides methodological support. In the Netherlands, there are different media labs in close relationship with the city and its different communities. For example, **Waag Futurelab** in Amsterdam, **V2 Lab for the Unstable Media** and **Worm** in Rotterdam.

The **Van Abbemuseum** has focused on sensory-cognitive aspects and different mental conditions. The museum has a program for people with reduced vision and held an international competition to work with the psychiatric hospital in the city of Eindhoven. **ARIAS** (Platform for Research Through the Arts and Sciences), based in Amsterdam, is dedicated to financing artistic research and advises post-MA researchers for doctoral degrees in their network, aimed at artist-researchers. Worth to mention the **Zone2Sourse, and TAAK Projects.**

At the Russian Federation, the **LABORATORIA Art&Science Foundation** in Moscow, is an exhibition space for artistic projects carried out under a non-profit organisation created by Daria Parhomenko in 2008.

Art&Science Node (ASN) in Berlin, is a network focused on the creation and synergy between science, art and technology that organises exhibitions and discussion panels. It has a residency system and conducts workshops of various kinds. Also in Berlin, the **State Studio exhibition laboratory**, founded in 2014 by physicist Dr. Christian Rauch, offers an open program to discuss works based on current trends in science, society and innovation, as a platform for collaboration between artists, scientists and the interested public. **INHABIT** is an Artists-in-Residence Program of the Max Planck Institute for Empirical Aesthetics, based in Frankfurt, where artists are invited to create in collaboration with the Institute's scientists and researchers.

The European Commission's Horizon 2020 Research and Innovation Programme, launched the **S+T+ARTS** (Science+Technology+ARTS) initiative. In what was called the Vienna Declaration, a series of associations belonging to the arts (performance, theatre, music and visual arts) came together to validate research programs considering artistic practices.

The **SOLU/Bioart Society**, based in Helsinki since 2008, develops residencies and publications among other activities with an emphasis in biology, ecology and the life sciences in general.

[1] For a complete evaluation of the program, see https://wellcome.org/sites/default/files/wtx057228_0.pdf.

The Arts & Sciences: Indisciplined Knowledge Program, from The Institute for Advanced Study Mediterranean (ImèRA), University of Marsella, France, is addressed to researchers working on science as well as with art, through situated research.

The **Institute for Research in Biomedicine IRB**, Barcelona, Spain, Artist in Residence Program seeks to connect interdisciplinary interests with a broad spectrum of research, scientific information and infrastructure. Worth mentioning the **Medialab-Matadero Madrid** (ex Media Lab Prado), **Laboral Art**, Industrial creation Centre in Gijon, also in Spain.

Japan, China and South Korea

In Japan, the best-known institution in the art and science integration is the **NTT Intercommunication Center**, created in Tokyo in 1990. This space establishes a dialogue between science, technology and artistic culture and serves as a centre for information and exchange between artists and scientists worldwide. Another space that matches these interests is the one created in 1996 under the name of **IAMAS, The Institute of Advanced Media Arts and Sciences**. This centre, located in the Gifú prefecture, produces scientific publications through the *Journal of the Institute of Advanced Media Arts and Sciences*, created by the same institution.

In the city of Matsudo, various initiatives have been developed with the aim of creating a city brand based on culture, where creators and artists play an active role. As part of these efforts, this city organised **Matsudo International Science Art Festival** an international event that connects science, art and nature, with activities such as exhibitions, workshops and conversations. At the **KYOTO STEAM Festival**, media artists, companies and research institutions, from traditional industries to cutting-edge research and universities, work and present together. Finally, it is also worth mentioning a series of exhibitions called **Art is Science at Karuizawa New Art Museum**. **BioClub** is a Bio Community Space located at FabCafe MTRL 2F in Shibuya, Tokyo.

In the People's Republic of China, some activities have been carried out around art and science. In 2020, the fifth version of the **International Art and Science Works Exhibition** was held in the **National Museum of China**, where both the exhibition and the symposium focused on artificial intelligence. Nine artworks on the theme were awarded and received the 2019 Tsinghua University Wu Guanzhong Prize for Artistic and Scientific Innovation. The National Museum of China also has a scientific research group focused on the traditional art of the different dynasties.

The **SJTU TD Science and Art Lecture Fund**, Shanghai Jiao Tong University LEE, was established in 2013 to promote and carry forward the spirit of the integration of science and art. This fund organises a Science and Art Works Competition each year to encourage the creation and exhibition of artworks based on scientific themes from around the world. Until the year 2023, nine competitions and exhibitions have been successfully held. The scientific topics involved "Metamaterial" (2022), "Liquid Crystal" (2021), "Particles of Light" (2019), among others.

In South Korea there have been a series of initiatives that link public interest with the technology industry. **The National Museum of Modern and Contemporary Art** in Seoul has held a series of art-science convergence exhibitions, such as *The Aleph Project* (2013), *Super Nature* (2014–15), and *Robot Essay* (2015). Related projects include *Project Daejeon 2014: The Brain*, in Daedeok Science City, which grafted brain science into art and the work of private organisations such **Art Center Nabi**, which continually introduce new media or technology-based artworks. Worth mentioning the **IBS Annual Art in Science Contest** on artistic images encountered by the scientists on their research.

North America

From an academic and outreach standpoint, the **Center for Science, Technology, Medicine, and Society (CSTMS)**, in collaboration with the **Arts Research Center (ARC) at UC Berkeley**, hosts an artists-in-residence program fostering exchanges between art, design, technology, science, and engineering. This research centre promotes rigorous interdisciplinary research with the conviction that the pressing problems of our time are simultaneously scientific and social, technological and political, ethical and economic, questioning the role of art in this vision. It provides a reflective space where scholars, curators, and artistic and civic leaders from various disciplines can meet and learn from each other.

The University of Illinois Urbana-Champaign held, in 2021, the **Art Science Festival** which was organised by Illinois Physics, specifically The Grainger College of Engineering. A group of physicists displayed art-related work, ranging from illustrations and theatre plays to computer models and sound compositions.

An important residency program to note is **6&6 Artists & Scientists**, a collaboration between artists and scientists to explore the patterns and processes of the Sonoran Desert and the Gulf of California. Los Angeles-based project called **SUPERCOLLIDER**, is a group of young artists who believe in a future where art, science and technology inspire social and environmental responsibility.

From an expository point of view, there are several universities that delve into the subject of art science. One of them was the exhibition organised by the Universities of Pittsburgh and NC State University, called ***Art's Work in the Age of Biotechnology: Shaping Our Genetic Futures***, during the years 2019–2020. This exhibition tries to answer questions around What do we want from biotechnologies? Who is biotechnology for and who decides it? Also noteworthy is the work being done in the Arizona desert, **The Tumamoc Arts and Science Program** at the Desert Laboratory of the University of Arizona, where lectures, workshops and courses are held, together with a transdisciplinary group, the **Art and Science team**.

It is also relevant to mention the **Scientific Delirium Madness**, a collaborative residence of six artists and six scientists held by **Leonardo/International Society for Arts, Sciences and Technology (ISAST)** and the **Djerassi Artists-in-Residence Program**. The goal of the project is to explore and expand the connection between the creativity of scientists and that of artists. **Coalesce: Center for Biological Arts**, from the University of Buffalo, USA, is a laboratory studio dedicated to creative approaches using life science tools and technologies.

The Fermilab laboratory, dedicated to particle physics research based in Chicago, with its **Guest Artist and Guest Composer program**, aimed to interact with scientists, learn about their research, and see how it connects to society.

In Canada, the **SNOLAB** scientific program studies extremely rare physical phenomena focusing in astroparticle physics, specifically neutrino and dark matter. Its exceptional location at the deep underground in the operational Vale Creighton nickel mine, near Sudbury, Ontario, is also suitable for biology and geology experiments. In 2020, the lab held its first transdisciplinary artistic residency **Dift**, which explores the search for dark matter from an artistic perspective in collaboration with the Agnes Etherington Art Center and the McDonald Institute. Also in Canada, The **Biophilium** school for artists encourages artists to appreciate and engage with wildlife.

The **Center for Science and Imagination** in Arizona State University brings together artists, authors, and educators with scientists, technologists, policy thinkers, and community members. They publish, lead educational initiatives, public events and interdisciplinary research. In this line is also the **Art|Sci Centre** from the University of California, Los Angeles, promoting scientific innovation research and programs on social, ethical and environmental issues, as part of a world-wide network of artists and scientists. In the **SVA Bio Art Lab**, interdisciplinary residencies take place in New York City's Chelsea gallery district where participants are introduced, among other subjects, to biological arts.

There are several examples with an outreach focus: **Interalia Magazine** is an online publication dedicated to the interactions between the arts, sciences and consciousness. **SciArt Initiative**, is a platform that focuses on art based on science and technology, interdisciplinary collaboration and education with a residency program called **The Bridge: Experiments in Science & Art**. It is a four-month virtual residency, open at the international level for professionals in the arts and sciences. Finally, the Artist-at-Sea Program at the Schmidt Ocean Institute, a private non-profit operating foundation, invites artists to work inspired from the research, scientists and crew aboard their research ship Falkor.

Australia

In the city of Perth is **SymbioticA,** at the University of Western Australia. Was the first research laboratory to receive artists and researchers working in wet biology or wetware practices, as they call it. It also hosts residents, workshops, exhibitions and symposiums. SymbioticA emphasises experimental practice with an informed critique of ethical and cultural issues related to the manipulation of the living. **Experimenta** is an organization presenting works in the intersection between art, science and technology since its inception in 1986. Their exhibiton, ***Experimenta Life Forms,*** held contemporary artists explore the notions and definitions of life, at a time when technological changes and new research discoveries make these definitions increasingly difficult to specify. Are our definitions of life changing due to new scientific discoveries? How do First Nations epistemologies influence ways of thinking and understanding life? How are notions of our place in the web of life changing now, that research is identifying sentience in animals, plant life, and perhaps soon in our machines?

A unique example of art-science is **The Big Anxiety Festival Project**, organised by the National Institute for Experimental Arts of the University of New South Wales, and the Black Dog Institute. Spanning diverse experiences across the broad spectrum of mental health, it examines what makes us feel anxious as individuals and social groups, from how to care for people to the fears about the future.

Finally, the **University of New South Wales Arts Scholarship Program** aims to increase Australian and international knowledge and appreciation of Antarctica, the Sub-Antarctic, the Southern Ocean and the Australian Antarctic Program. The Arts Fellowship furthers these goals through the work of individuals gifted in communicating through various media, including (but not limited to) the visual arts, filmmaking, acting, writing, and music.

6.3 Postgraduate Programs

Several post graduate programs focused on art-science integration have emerged in many places around the world. Closer to the approach presented in this book and with a strong commitment with interdisciplinarity, some of these programs are listed below:

M.Sc. in Art & Science. Art Science Centre. ITMO University, San Petersburgo, Russia.
Doctorate program in Design, Art and Science. University of Bogotá Jorge Tadeo. Lozano, Colombia.
Master of Arts in Arts, Science, and Health, ASH MA. University of Zagreb, Croatia.
Art & Science Master's program. University of Applied Arts. Vienna, Austria.
Art & Science Ph.D. program. The Art x Science School for Transformation. Johannes Kepler University Linz and University of Applied Arts Vienna, Austria.
MA Art and Science from the University of the Arts London, UK.
Masters degree. Arts, Computation and Engineering program. University of California Irvine, USA.

Master and PhD programs. Media Arts and Sciences (MAS). MIT Media Lab. Massachusetts, USA.

MS and Ph.D. programs. Arts, Media and Engineering program. Arizona State University, USA.

Master in Art & Social Practice Program. Portland State University, USA.

Master of Biological Arts. Symbiotica. University of Western Australia.

Masters and Ph.D. Program in Media Arts and Sciences. The Institute of Advanced Media Arts and Science (IAMAS). Gifu, Japan.

Ecology Futures Master programme. Master Institute of Visual Cultures. St. Joost School of Art & Design. School of Art Design. 's-Hertogenbosch, the Netherlands.

Diriyah Art Futures Emerging New Media Artist Programme. Ministry of Culture, Riyadh Kingdom of Saudi Arabia.

Master ArtScience. ArtScience Interfaculty, The Royal Academy of Art (KABK) and the Royal Conservatoire. University of the Arts The Hague, the Netherlands.

Interestingly, in 2021 a postdoctoral position in ecology was opened at the Academy of Fine Arts at the University of Helsinki.

Bibliography

Scott J (2006) *Artists in Lab. Process of Inquiry.* Springer; Auflage.

MIX
Papier aus verantwortungsvollen Quellen
Paper from responsible sources
FSC® C105338

If you have any concerns about our products,
you can contact us on
ProductSafety@springernature.com

In case Publisher is established outside the EU,
the EU authorized representative is:
Springer Nature Customer Service Center GmbH
Europaplatz 3, 69115 Heidelberg, Germany

Printed by Libri Plureos GmbH
in Hamburg, Germany